LIZARDS AND CROCODILIANS OF THE SOUTHEAST

Lizards
& Crocodilians

OF THE SOUTHEAST

by Whit Gibbons, Judy Greene, and Tony Mills

The University of Georgia Press   Athens and London

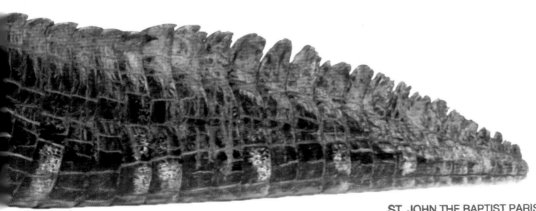

© 2009 by the University of Georgia Press

Athens, Georgia 30602

www.ugapress.org

Designed by Mindy Basinger Hill

Set in 10/15 Scala

Printed and bound by Everbest

through Four Colour Print Group

The paper in this book meets the guidelines for

permanence and durability of the Committee on

Production Guidelines for Book Longevity of the

Council on Library Resources.

Printed in China

13 12 11 10 09 P 5 4 3 2 1

Library of Congress Cataloging-in-Publication Data

Gibbons, J. Whitfield, 1939–

Lizards and crocodilians of the southeast / by Whit Gib-

bons, Judy Greene, and Tony Mills.

 p. cm. — (A wormsloe foundation nature book)

Includes bibliographical references and index.

ISBN-13: 978-0-8203-3158-4 (pbk. : alk. paper)

ISBN-10: 0-8203-3158-9 (pbk. : alk. paper)

1. Lizards — Southern States. 2. Crocodilians —

Southern States. I. Greene, Judy. II. Mills, Tony, 1962–

III. Title. IV. Series.

QL666.L2G53 2009

597.950975 — dc22   2009001352

British Library Cataloging-in-Publication Data available

Contents

INTRODUCTION

Why Lizards and Crocodilians?   1

Defining the Southeast   3

Exotic Lizards and Crocodilians in the Southeast   5

Naming Lizards and Crocodilians   6

ALL ABOUT LIZARDS

Lizard Biodiversity   11

Families of Native and Exotic Lizards Found in the Southeast   13

General Biology, Ecology, and Behavior of Lizards   23

How to Identify Lizards   36

ALL ABOUT CROCODILIANS

Crocodilian Biodiversity   41

General Biology, Ecology, and Behavior of Crocodilians   42

SPECIES ACCOUNTS

Organization and Order of Species Accounts   55

Native Lizards   59

Introduced Lizards   118

Native Crocodilians   189

Introduced Crocodilian   203

PEOPLE AND LIZARDS AND CROCODILIANS
Studying Lizards and Crocodilians   209
Keeping Lizards and Crocodilians as Pets   213
Conservation of Lizards and Crocodilians   216

**WHAT LIZARDS AND CROCODILIANS
ARE FOUND IN YOUR STATE?**   220

GLOSSARY   223

FURTHER READING   225

ACKNOWLEDGMENTS   227

CREDITS   229

INDEX OF SCIENTIFIC NAMES   231

INDEX OF COMMON NAMES   233

LIZARDS AND CROCODILIANS OF THE SOUTHEAST

Introduction

WHY LIZARDS AND CROCODILIANS?

Everyone is familiar with lizards and crocodilians, and many people find them fascinating and even attractive. Anyone living in the southeastern United States who spends much time outdoors is likely to encounter one or more species of lizards, even in many urban areas. Alligators, the most common species of U.S. crocodilian, have once again become abundant in many aquatic habitats within their natural range. Enforced protection originating with the Endangered Species Act and continuing education efforts through federal and state wildlife programs have brought alligators back from severe population declines that many feared were approaching extinction for the species. Americans are showing an increasing interest in maintaining the environmental integrity of our remaining natural areas, and this has led to a desire to understand our native wildlife. Lizards and crocodilians offer opportunities for exploration and new "firsthand" experiences in the Southeast. Our intent with this book is to teach people about these exciting creatures, many of which are literally in backyards in many areas. We hope that knowledge about the behavior and ecology of these animals will increase people's enjoyment of and appreciation for natural areas of the Southeast.

American alligators remain common residents in many areas of the Southeast as a consequence of the U.S. Endangered Species Act and continuing efforts by state wildlife programs.

A male green iguana in orange breeding color courts a female prior to mating.

An African rainbow lizard (above) from Kenya, one of the thirty-eight species from other countries that became established in the Southeast, mostly in southern Florida, by the early twenty-first century

Broad-headed skinks (right) can be found in suburban neighborhoods throughout much of the Southeast.

The traditionally recognized major groups of reptiles native to the United States are snakes, turtles, lizards, and crocodilians. Two earlier books in this series cover the more than 50 species of snakes and 40 species of turtles found in the Southeast. Because of the small number of naturally occurring species of lizards and crocodilians in the region, we have combined the 20 species of native lizards and crocodilians in a single book. In addition, we provide brief accounts of 2 species of lizards native to the U.S. Southwest that have been introduced into the Southeast and 39 exotic species from other countries (38 lizards and 1 crocodilian) that have become established here; most are in southern Florida, but some can be found as far north as Kentucky.

The small, land-dwelling lizards such as green anoles and little brown skinks have several biological traits in common with the enormous, water-inhabiting crocodilians. An elongate body with four legs and a long tail give typical lizards and crocodilians a similar body shape. In fact, the name for the country's best-known crocodilian, the alligator, is believed to be derived from the Spanish *el lagarto*, meaning "the lizard," presumably because of that superficial similarity. Both lizards and crocodilians have bodies armored with scales — a general reptilian trait — both have clawed feet (except for a few species of southeastern lizards that are legless), and both have a large mouth in relation to their body size. The two groups differ significantly in many anatomical features, however, including the structure of the heart, basic skull structure, and dentition (teeth). In fact, reptile experts believe that lizards and crocodilians have very different evolutionary origins. Lizards are close relatives of snakes, while crocodilians, oddly enough, are more closely related to birds than to lizards.

Lizards and crocodilians have the lowest biodiversity among the major groups of reptiles and amphibians found in the Southeast, and both groups are represented by far fewer species here than they are in some other regions of the world. The tropics are particularly rich in lizard and crocodilian species.

Did you know?

The largest lizards ever known were the giant mosasaurs, now extinct, that lived in the oceans during the Cretaceous and reached lengths in excess of 50 feet.

DEFINING THE SOUTHEAST

Land areas of the country can be partitioned on the basis of regional culture, political boundaries, or geological and environmental features, and the Southeast has been defined in all those ways. The primary factors determining lizard and crocodilian geographical distributions in the Southeast are climatic temperatures and rainfall. Most of the lizards native to the United States are restricted to hot, arid regions of the Southwest, whereas

Like this adult collared lizard, most native U.S. lizards occur primarily in hot, arid regions of the Southwest.

The U.S. geographic range of the American crocodile is restricted to southern Florida.

the range of native crocodilians is limited to warm, humid habitats of the Atlantic and Gulf coastal plains. To maintain the practicality of using states as the basic units for describing geographic ranges of the species, we have chosen to define the Southeast as Alabama, Florida, Georgia, Kentucky, Louisiana, Mississippi, North Carolina, South Carolina, Tennessee, and Virginia. The defined region has an identifiable assemblage of southeastern lizards. Moving the boundaries farther west into Texas or Arkansas would include an array of species typically associated with drier southwestern habitats. The prescribed area also encompasses the vast majority of the natural range of the American alligator and the total U.S. range of the American crocodile.

The brown basilisk is one of Florida's exotic lizards from tropical America.

EXOTIC LIZARDS AND CROCODILIANS IN THE SOUTHEAST

Ironically, although lizard and crocodilian species native to the Southeast occur in lower numbers than do the other major groups of reptiles and amphibians, they constitute a greater proportion of the exotic species that have been introduced and have become established. The number of introduced lizard species, 40, is more than double the number of native ones. In fact, more species of introduced lizards thrive in the Southeast than do introduced exotic species of all other reptiles and amphibians combined. Likewise, the introduced spectacled caiman represents one-third of the crocodilian species living in the United States.

With a few exceptions (e.g., Mediterranean gecko, brown anole), lizards introduced from other countries are restricted to Florida, especially the extreme southern portion of the state. At least two U.S. species (collared lizard and Texas horned lizard) not native to the Southeast have been reported from southeastern states outside Florida.

The first exotic lizard species (brown anole) recorded as an inhabitant of Florida was reported in the 1800s; additional species were noted in the early part of the twentieth century. Their species diversity and potential to become established residents did not become evident, however, until the last quarter of the century. By the early 2000s, more than 30 species of exotic lizards had been documented from the lower half of the Florida peninsula. Many were in suburban neighborhoods, state parks, and de-

The brown anole was the first exotic lizard species to be recorded, in the 1800s, as an inhabitant of Florida.

A spiny head crest is characteristic of many lizard species such as this Jamaican giant anole.

veloped urban habitats, and some were becoming noted as predators or competitors of native lizards.

The single most common origin of Florida's exotic lizards is the West Indies (13 species), including Cuba, Jamaica, and Hispaniola. Appreciable numbers of species (6–10) originated in Africa, Asia, and mainland areas of tropical America. How and when an exotic species arrived in Florida may not be known, but the pet trade is associated with the introduction of many. The intentional release of pets that owners no longer want, escapees from pet stores, and damaged shipping containers at airports and commercial docks have probably all been responsible for lizard introductions in one place or another. Damage to pet stores, animal dealer locations, and zoos by hurricanes or tropical storms has probably resulted in some escapes of exotic lizards as well. Some species have been introduced, and have become established, at more than one location.

NAMING LIZARDS AND CROCODILIANS

Herpetologists (scientists who study reptiles and amphibians) follow strict taxonomic rules to determine the unique scientific name that is given to a particular lizard to distinguish it from other species. Like all plants and animals, each type of lizard and crocodilian has a two-part scientific name comprising a genus (plural "genera") name (e.g., *Anolis*), which is always capitalized and italicized, and a specific designation (e.g., *carolinensis*) that is also italicized but not capitalized and is referred to as the specific epithet. Hence, the scientific name of the green anole is *Anolis carolinensis*. Some species are further partitioned into subspecies—morphologically distinct races that occupy different geographic ranges—resulting in a third taxonomic name, which is also not capitalized (e.g., *Anolis carolinensis seminolus*).

Herpetologists place closely related species within the same genus. All species of anoles, for example, belong to the genus *Anolis,* and all of the species of glass lizards are in the genus *Ophisaurus.* Closely related genera are placed within the same family. For example, the Florida reef gecko, *Sphaerodactylus notatus,* is in the family Gekkonidae along with geckos in the genera *Hemidactylus, Gekko, Chondrodactylus,* and many others. The nested relationships of species, genus, and family parallel the relationships of city, county, and state.

Common or vernacular names do not follow any established taxonomic rules and are typically based on what people of a particular region call a species. Common names thus vary across geographic regions and within different cultural groups. The names for U.S. crocodilians (alligators, crocodiles, and caimans) are consistent throughout the country, but the names for some lizards vary geographically. For example, most herpetologists and much of the public call *Anolis carolinensis* a green anole, but the name chameleon is still used in many parts of the species' range because of the anole's ability to change its color. The U.S. species in the genus *Ophisaurus* are variously called glass lizards, glass snakes, legless lizards, and joint snakes. No particular common name is more correct than another, as long as it is clear what species is being referred to.

Green anoles and many other lizards spend much of their time basking in vegetation with partial shade and sun, a behavior that helps regulate body temperature.

Taxonomic Controversies

Late-twentieth-century technological advances in the biological sciences, such as the ability to compare DNA sequences, have allowed taxonomists to reevaluate ancestral relationships among lizards and many other organismal groups. As a consequence, classification changes at the genus and even family levels have been proposed for many species. Even some of the relationships among groups at higher levels of classifications, traditionally known as orders and classes, have been challenged. We now know that crocodilians are more closely related to birds than to reptiles, for example, and that turtles are only distantly related to any of the classical reptile lines. Another radical change based on genetic evidence combines all snakes and lizards into a single group in which some species of lizards are viewed as more closely related to snakes than to other lizards.

Alligators and other crocodilians are more closely related to birds than to lizards, snakes, or turtles.

When their mouths are closed, many crocodilians such as this alligator appear to be smiling.

The scientific community does not always accept such reinterpretations of relationships, and often the original, traditional name for a species remains unchanged even after advanced analysis. For the purposes of this book, the most important point is that the authors and the readers know exactly which animal is being discussed. Whether the brown anole is placed in the genus *Anolis* or *Norops,* whether all of the anoles are placed in the family Iguanidae or Polychrotidae, or whether the six-lined racerunner is placed in the genus *Cnemidophorus* or *Aspidoscelis* does not change the behavior, ecology, or general biology of the animal. However, to minimize confusion that may be created by future advances in taxonomy, we mention such controversies and use the scientific names that we believe are or will become most widely used. All of the scientific names will certainly be familiar to most herpetologists.

Early in the twenty-first century, lizard taxonomists proposed changing the genus name of the six-lined racerunner of the Southeast (shown here) and the whiptail lizard of the Southwest from *Cnemidophorus* to *Aspidoscelis.*

A Komodo dragon walks the beach on the island of Komodo. These are the largest living lizards. Some males reach a total length of nearly 10 feet.

All About Lizards

LIZARD BIODIVERSITY

Although we traditionally think of lizards as small, inoffensive creatures that climb on vegetation or walls, they display great diversity in behavior, ecology, and even morphology on a global scale. Among the historically recognized groups of reptiles and amphibians, lizards are by far the most numerous in terms of family groups and species, with more than two dozen families and more than 4,000 named species. The group includes not only the expected four-legged species but also two-legged and legless ones. When it comes to size, lizards are also quite diverse. The adults of different species range from the tiny reef gecko of Florida, less than 3 inches long (including the tail), to the enormous Komodo dragon, which can be almost 10 feet long. Most lizards are harmless to humans, although many will try to bite if handled, usually resulting in at most a mildly painful pinch. Some, however, including the Gila monster and beaded lizards of the U.S. Southwest and Mexico, are actually venomous, and their bites can be excruciating and even fatal.

The Southeast is the natural home of 18 species, eight genera, and six families of native lizards, of which three (Amphisbaenidae [wormlizards], Teiidae [racerunners], and Gekkonidae [geckos]) are represented by only a

The broad-headed skink belongs to the family of lizards having the most species in the Southeast and the world.

Green anoles often hibernate in large numbers, emerging during warm weather. Five individuals can be seen on the outside of a stump where they spent the colder months.

A population of veiled chameleons that lives in southern Florida belongs to a family of lizards native to Yemen and Saudi Arabia and not found naturally in the Western Hemisphere.

single native species. The family having the most species (8) is the Scincidae (skinks). The remaining families—Iguanidae (anoles and fence lizards) and Anguidae (glass lizards)—have 3 and 4 species, respectively, in the Southeast. Almost half of the southeastern species of lizards are endemic; that is, confined to this region.

At least two species of lizards native to the American Southwest have been reported in southeastern states. The Texas horned lizard has unquestionably established breeding populations that have persisted for many years in several locations, and the collared lizard has been found in Louisiana more than 100 miles east of the nearest known population in Texas. Established populations had not been documented anywhere in Louisiana by the early 2000s, but the species is included here because other field guides mention its possible presence in Louisiana.

Although the species diversity of southeastern lizards is low compared with the U.S. Southwest, a remarkable biological phenomenon has become apparent in subtropical southern Florida: more than 30 species of lizards from other countries have become established. More of these pioneering species are likely to follow. The documented exotic invaders represent four families that are also native to the United States: 14 species from the family Iguanidae, 12 Gekkonidae, 4 Teiidae, and 1 Scincidae. Oddly enough, although more species of the latter family are found in the Southeast and the world than species of any other family of lizards, only 1 species of Scincidae has been introduced successfully into the Southeast. In addition, representatives of three families that do not occur naturally in the Western Hemisphere—Agamidae (agamid or Old World lizards), Varanidae (monitors), and Chamaeleonidae (true chameleons)—now thrive in southern Florida.

Diversity can also be viewed in terms of how many different habitats a group occupies. Southeastern lizards can be found in an assortment of natural and modified habitats. Some prefer sandy terrain, for example, while others favor hardwood or pine forests, or even suburban areas. No lizard native to the United States is found exclusively in or around fresh or salt water.

FAMILIES OF NATIVE AND EXOTIC LIZARDS FOUND IN THE SOUTHEAST

Amphisbaenidae (Wormlizards)

Some herpetologists consider members of this poorly studied group, known as wormlizards or amphisbaenians, to be very distant relatives of other lizards. In fact, from an evolutionary perspective, snakes are as closely related to lizards as wormlizards are. Taxonomic authorities place all or most species of the 24 genera and 140 species of amphisbaenians into a single family, but 10 species are sometimes separated into three additional families because of their uncertain evolutionary heritage. Amphisbaenians have their greatest biodiversity and are most geographically widespread in Africa and South America, and a few species occur in Cuba, southern California, and Turkey. The single southeastern species, the Florida wormlizard, is sometimes placed in its own family (Rhineuridae) instead of in the family Amphisbaenidae. Most of the wormlizard species that have been studied lay eggs, although a few are livebearers.

Wormlizards look like large earthworms. Visible rings encircle the body, and with the exception of 3 species found in California and Mexico that have front legs, they are legless. Amphisbaenians do not have ear openings as the true lizards do. Adults of the largest species reach lengths in excess of 2 feet, but most are only about a foot long and the smallest species are less than 6 inches long.

A Florida wormlizard looks more like an earthworm than a lizard. They have no legs, no ear openings, and no visible eyes.

Green iguanas (right) are one of the largest and most impressive of the introduced lizards that now inhabit southern Florida.

Some herpetologists consider fence lizards of the Southeast (upper left) to be in the family Iguanidae, whereas others consider them phylogenetically distinctive enough to be in their own family, Phrynosomatidae.

Iguanidae (Anoles and Their Relatives)

This family includes as many as eight groups of lizards that some herpetologists consider to be subfamilies (among them are Crotaphytinae, Corytophaninae, Iguaninae, Phrynosomatinae, Polychrinae, and Tropidurinae). Other lizard biologists place these groups in separate families within a "superfamily." Regardless of the taxonomic scheme used, anoles, iguanas, basilisks, and collared lizards are all recognized as distinct groups; horned lizards and fence lizards are lumped together in the same group. The family Iguanidae includes under its umbrella 44 genera and more than 650 species, of which the anoles constitute about 200. Only 3 iguanid species are southeastern natives, but 16 exotic iguanids have become established — more than any other family. Among them are 2 species that have been introduced from the western United States. Most iguanids are native to the Americas and occur in climates from the temperate zones through the tropics. The anoles are especially prevalent on islands throughout the West Indies. The marine iguanas are found on the Galápagos Islands. Old World representatives occur in Madagascar and the Fiji Islands, but there are no native iguanids in mainland Eurasia, Africa, and Australia.

All iguanids have four limbs, and all but a few reproduce by laying eggs. They occupy a wide range of habitats. Some are strictly arboreal, others occupy deserts, and some are associated with rocky terrain. Most of the

Although many of the exotic anoles have green body coloration like the native green anole, body size can often be used to differentiate between species, as with the large size of the knight anole (shown here).

species are insect eaters, but a few—such as green iguanas—are primarily herbivores. Most reach a total length of only a few inches, but some of the iguanas can be more than 6 feet long.

Teiidae (Whiptail Lizards and Tegus)

The nine genera and 120 or so species of whiptail lizards and tegus occur throughout most of temperate and tropical North and South America, and nowhere else. Most are terrestrial rather than arboreal or fossorial, and they occupy desert habitats as well as forests, coastal areas, and grasslands. A few South American species are actually semiaquatic and spend most of their time in or around water. All species for which reproductive information is known lay eggs, and several species are parthenogenetic, including some found in the U.S. Southwest.

Most members in the family Teiidae have four limbs and a long tail that will break and regenerate if a predator grabs it, although the regrown tail tends to be much shorter and blunter than the original. The body scales on the back and sides are tiny and have a granular appearance; the belly scales are noticeably wider. Many of the whiptails are less than a foot long, but some of the tegus exceed 3 feet. Only one species in the family, the six-lined racerunner, is native to the Southeast.

The giant ameiva is one of the South American whiptail lizards with established populations in southern Florida.

Eastern glass lizards are common on some barrier islands. These are from Kiawah Island, South Carolina.

Southeastern glass lizards and some other members of the family Anguidae have a side groove that begins behind the head and runs the length of the body.

Anguidae (Glass Lizards)

The legless glass lizards of the Southeast belong to a family with 13 genera and more than 100 species, among them the alligator lizards and California legless lizards of western North America. The family is well represented in the Americas, Europe, and Asia.

Glass lizards characteristically have heavy protective scales as well as osteoderms (small, bony protective plates embedded beneath the outer body scales). Many members of the family, including the 4 species of southeastern glass lizards (genus *Ophisaurus*), have a fold that runs the length of the body on each side. This fold—sometimes called a groove—gives flexibility to the otherwise rather rigid body, allowing it to expand when the animal breathes or eats a large food item, or when females are laden with eggs. Like most other species of lizards, the glass lizards can lose part of the tail to avoid predation. The length of the intact tail varies from one species to the next. Most species have fracture planes in the vertebrae that will break readily when the animal is attacked by a predator—or handled by a person. The tail of southeastern glass lizards is typically two-thirds or more of the total length and will regenerate if broken. The regrown tail is uniformly brown with a pointed scale at the end and never quite reaches the length of the original. Regenerated tails are cartilaginous rather than bony and will not fracture as readily. Most species in the family, including all of the southeastern glass lizards, lay eggs. Adults range in total length from less than a foot long to the 4 feet or more of the European glass lizard. Despite their superficial resemblance to snakes, the glass lizards all have eyelids and ear openings in the side of the head (both of which snakes lack). Also, because of their osteoderms, the legless lizards are more rigid and move much more stiffly than snakes.

Southeastern glass lizards are glossy, streamlined, legless lizards that casual observers often mistake for snakes. Unfortunately, people needlessly fearful of snakes often kill them before taking a closer look. One does have to look closely to see the eyelids and external ear openings. These lizards are the source of a southern myth about a "snake" that shatters into pieces when hit with a shovel. Some people believe that, in contrast to Humpty Dumpty, the pieces crawl back together later and re-form the "glass snake's" body. The origin of such folklore is understandable: a legless reptile with a glossy body that looks like a fragile piece of porcelain and can indeed shatter into multiple pieces, sometimes without even being touched! Such a dead "snake" scavenged overnight by some hungry wild creature would perpetuate the notion that the animal had reassembled itself and crawled away. All of the glass lizards of the Southeast are active during daytime and dusk and are unlikely to be out at night.

Scincidae (Skinks)

The skinks comprise the world's largest and most diverse family of lizards (120 genera and more than 1,400 species, approximately 31 percent of all lizards). This family is also better represented by native species than any other lizard family in the Southeast (2 genera and 8 species, or about 42 percent of our native lizard species). Skinks occur on every warm continent and many oceanic islands with a warm climate, and they occupy a variety of habitats, primarily terrestrial ones. Some species of skinks are arboreal and some are fossorial, and both kinds are found in the Southeast.

The family includes both oviparous (egg layers) and viviparous (livebearers) species, but all southeastern species lay eggs. The U.S. species are relatively small compared with the large sleepy lizards and blue-tongued skinks of Australia. Most skinks have legs, although some have greatly reduced limbs and some are limbless. All of the southeastern species have four relatively small limbs. Most of the world's skinks, and all those in the United States, have smooth, shiny scales covering the body. Like the glass lizards, skinks have osteoderms—small, bony protective plates

Most species of southeastern skinks have body stripes and brightly colored tails.

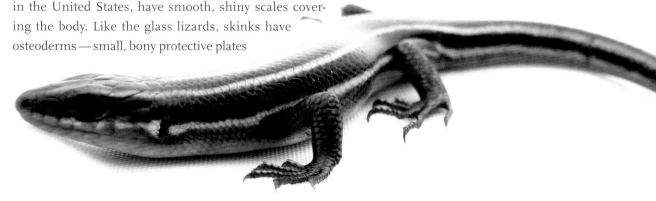

Juvenile skinks of some species, including the coal skink (shown here), differ from the adults in appearance.

embedded beneath the external body scales—and most skinks are capable of losing their tail to escape predators. Broad-headed skinks, and probably other species of skinks, have a vomeronasal system that allows them to detect subtle chemical cues in the environment. Skinks use this chemosensory system to discriminate among prey types and to determine if another skink is of the same species, its sex, and its reproductive status.

Gekkonidae (Geckos)

Of the more than 1,000 species of geckos around the globe, the Mediterranean gecko is perhaps the most widespread. The introduced species has been reported to occur in every southern state from South Carolina to California.

The more than 100 genera and nearly 1,000 species of geckos are distributed throughout all or most of Africa, Europe, Asia, South America, and Australia, and on numerous tropical islands. Only a few native species occur in the United States—in south Florida and the Southwest—but at least a dozen species have been introduced, and some have become established in Florida. The Mediterranean gecko has breeding populations in at least seven southeastern states.

As would be expected of a group of lizards with so many species, extensive variability is apparent in morphological, physiological, and ecological traits. For example, most geckos are active at night, but a few are diurnal; some have four legs whereas others have hind limbs only. Most have no eyelids, and some have vertical pupils. Most geckos lay eggs, but a few are livebearers. Some are parthenogenetic, including the Indo-Pacific gecko that has been introduced into Florida; males are unknown in this species. A complete clutch of eggs for a gecko is only one or two, depending on the species. A few of the largest geckos living today reach a total length of 2 feet or more, but most species are much smaller. Many species of geckos can walk easily up vertical walls of wood, rock, or even glass by virtue of tiny hairlike structures on their feet. These structures create a type of molecular force that acts as a "biological magnet" and allows them to adhere to any surface, natural or man-made.

Tokay geckos are one of the many species in the family Gekkonidae that have established populations in southern Florida. They will vigorously defend themselves from people or predators by delivering a powerful and painful bite.

Agamidae (Old World Lizards)

This Old World family of lizards comprising 52 genera and more than 400 species is widespread throughout much of Africa, Asia, and Australia. Ecologically, many of the agamid species are counterparts of the iguanids of the New World. The majority are terrestrial or arboreal; most are carnivorous, but a few are herbivorous. All four limbs are usually well developed. The agamid species that have been studied are egg layers. Individuals of the largest species are more than 2 feet long, but most are much smaller.

This male red-headed agama from Mali, like other species in this Old World family, superficially resembles species in the family Iguanidae from the New World.

Most true chameleons from the Old World, like this veiled chameleon, live most of their lives in aboveground vegetation.

Chamaeleonidae (True Chameleons)

These bizarre lizards are familiar to almost everyone because of their distinctive body shape and their ability to change color to match their background. The 130 species, belonging to four genera, are distributed throughout much of Africa, Madagascar, and southern India. Traits of their peculiar anatomy include a vertically flattened body; a prehensile tail able to hold on to objects such as branches; a tongue that can be propelled forward to catch prey; and eyes that operate independently, one looking forward, for example, while the other looks behind. Most are arboreal and stay in trees or other vegetation for their whole life except for trips to the ground to lay eggs. Individuals of the smallest species of chameleons are tiny, reaching only a little over an inch in length, but the largest chameleons are more than 2 feet long. One introduced species occurs in the Southeast.

Lacertidae (Wall and Rock Lizards)

The more than 25 genera and 220 species in this family are known as wall lizards or rock lizards. Most are medium-sized, but a few exceed a foot in total length. Their general body form is similar to that of the whiptail lizards (family Teiidae) of the Americas. The lacertids are distributed throughout Europe, Africa, Asia, and the East Indies but are not native to Australia or the Americas. Their similarities to whiptail lizards extend to their use of terrestrial habitats that are characteristically arid and often rocky. They have a mostly insectivorous diet. Most species lay eggs, although some, like their American counterparts in the family Teiidae, are parthenogenetic.

The common wall lizard, of the family Lacertidae, is the only species of reptile that has become established in Kentucky but not in Florida. Old World lizards in this family resemble the New World whiptail lizards.

Varanidae (Monitors)

The largest lizards in the world—the goannas and monitors—are members of this family, which comprises only one genus with more than 40 species. Their physical characteristics include a long neck with a snakelike head and tongue, well-developed legs, and a thick body with a long tail that is not designed to break off. Some can stand erect on their back legs using the tail for support. The varanids are native to Africa, Asia, and Australia. Adults of the smallest species reach lengths of only 7 or 8 inches; several of the largest monitors get to be nearly 10 feet long and weigh 300 pounds.

Nile monitors, established in parts of Florida, are the largest lizard species with viable populations in the United States.

Courtship and territorial displays are common among many species of lizards, as seen in a native green anole (top) and an introduced brown anole (bottom).

GENERAL BIOLOGY, ECOLOGY, AND BEHAVIOR OF LIZARDS

Lizards are reptiles whose closest relatives are snakes. They characteristically have an elongated body and four legs, but the body shape varies widely—from the stout, sausage-like southwestern Gila monster to the slender, almost threadlike Australian flap-footed lizards to the humpbacked, gargoyle-like African chameleons. The legs may be large and sturdy or relatively small, and may number four, two, or none. Like typical reptiles, lizards have scales covering the entire body; these may be enlarged plates, rounded bumps, or granular beads. In a process known as ecdysis, lizards regularly shed patches of the outer layer of the skin that covers the scales. Most lizards have movable eyelids, although some of the geckos have only a transparent lens called the spectacle to protect their eyes.

All lizards have teeth designed for holding and manipulating their prey, or vegetation if they are herbivores. Some species have specialized teeth, such as the large molars of South American caiman lizards, which crush the snails on which they generally feed. Australian snake lizards that eat skinks have hinged teeth that can be inserted between the skink's protective bony plates. Many large monitor lizards have curved, serrated teeth perfect for ripping apart large vertebrate prey, and Gila monsters and beaded lizards have grooved teeth that direct the flow of venom produced by glands in the lower jaw.

Many species of lizards, such as this glass lizard, are visual hunters with excellent eyesight.

The internal anatomy and organ systems (e.g., cardiovascular, skeletal, muscular, digestive, reproductive, nervous) of lizards are generally similar to those of other vertebrates, with specializations related to their particular ecological and evolutionary history. Some lizards, for example, have a parietal (also called pineal) gland on top of the head. This "third eye" is sensitive to light and keeps the lizard's biological clock in synchrony with changing day lengths. Most lizards have external ear openings and can hear airborne sounds. Species whose males exhibit colorful patterns during courtship can presumably see color. Like snakes, lizards have a forked tongue and a well-developed combined sense of taste and smell. The tongue is flicked out to gather chemical odor particles from the surroundings, and then brought back inside and into contact with the Jacobson's organs in the roof of the lizard's mouth. The Jacobson's organs analyze and identify these chemical "samples." Most lizards have good vision, although a few burrowing forms, such as the Florida sand skink and wormlizards, have greatly reduced eyes or no eyes at all.

Individual lizards of some species, such as this green anole, can change color on all or parts of the body.

Activity and Locomotion

Many people might be surprised to learn that some lizards can dart overland at speeds faster than the average human can run. Six-lined racerunners traveling over sandy terrain are the fastest of our native species of lizards, but even legless glass lizards can undulate remarkably rapidly across their preferred substrate for short distances. Florida wormlizards and Florida sand skinks are unusual in being most active when burrowing below the surface of sandy soils.

The degree of activity of any particular lizard species is influenced by the time of day, season of the year, and recent or sometimes longer-term environmental conditions. In contrast to the many burrowing snakes that

Glass lizards, although legless, can move rapidly across the ground.

A hatchling six-lined race-runner begins hunting for its first meal soon after leaving the egg and is able to move rapidly across the ground in search of prey or to escape predators.

live most of their life underground, most lizards are active only above-ground. Almost all southeastern lizards are diurnal; that is, they move around during the daytime and are not active after dark. The reef gecko of Florida is active in the early evening, though, and most of the introduced geckos are active at night.

Although most lizards native to the Southeast are active during the daytime, many introduced geckos, such as this ocellated gecko, are primarily active at night.

Like all reptiles and other "cold-blooded" animals, lizards are affected by environmental temperatures and are more active during warm seasons than cold ones. During the coldest part of the winter, most native lizards of the Southeast brumate (hibernate), retreating to hideaways beneath tree bark; in rock crevices and underground burrows; or under logs, leaf litter, or other ground debris where they remain inactive until warmer temperatures return. Lizards in the southernmost regions of Florida and coastal areas of the southern Atlantic and Gulf coasts usually spend fewer days in brumation (hibernation) than do lizards in more northerly areas or in mountainous habitats.

Most species emerge from their cold-weather hiding places craving food at just the time when many kinds of insects begin to appear. Southeastern lizards are most active during the spring courtship period, when males of some species compete for females by combat or through behavioral displays. Snakes increase their activity in the fall as they move from summer habitats to hibernation sites, but no southeastern lizards migrate appreciable distances overland, and their activity in the fall is negligible.

On sunny days during cool weather, skinks, fence lizards, and anoles commonly bask on warm surfaces or in patches of sunshine to raise their body temperature. In most of the Southeast, at least some lizards are active on warm, sunny days in winter and therefore potentially can be seen in every month of the year. Green anoles are noticeably active in winter during spells of warm weather.

Basking in the sun on rocks or trees is the means by which most lizards, such as this brown anole, elevate their body temperatures.

Temperature Biology

Temperature influences most aspects of the biology of lizards—particularly the rate of growth, effectiveness at capturing prey, and ability to outrun or outmaneuver predators. Because the body temperature of lizards is determined primarily by the surrounding environmental temperatures, lizards are colloquially referred to as "cold-blooded." Scientists prefer the term *ectothermic,* which means that the body temperature is determined externally—outside the body—rather than being controlled internally, as is characteristic of "warm-blooded" (*endothermic*) animals. A lizard's body temperature tracks that of its immediate surroundings, rising or falling as the chosen locale warms or cools. Birds and mammals, including humans, are endotherms that maintain their body temperature through the production of internal body heat. The metabolism of most species of birds and mammals is typically high, and their body temperature varies by only a few degrees within a 24-hour period regardless of the surrounding environmental temperature. In contrast, the normal body temperature of some desert lizards can range from 93°F (34°C) to more than 102°F (39°C) on a daily basis. Forest-dwelling lizards may be active at temperatures below 79°F (26°C), but if given the opportunity to bask in the sun may elevate their body temperature above 95°F (35°C). All southeastern lizards can tolerate low temperatures approaching freezing, although introduced tropical species may be unable to tolerate sudden cold snaps; that is one reason why they are more common in southern Florida's subtropical climate than elsewhere in the Southeast.

Lizards actively adjust their body temperature by moving around to regulate their level of exposure to sun or shade and by choosing resting or hiding places that protect them from extreme temperatures. Being ectothermic can have disadvantages for a species, such as the inability to fend off predators effectively during cold weather, but ectothermy has biological advantages as well. Ectotherms typically have a low metabolic rate and thus require less energy than do birds or mammals. A mammal may need up to ten times as much food during a year as a lizard of similar size. The ability to function at a low metabolic rate for much of the year allows lizards to funnel a greater proportion of their energy intake into growth and reproduction.

Food and Feeding

Most lizards native to the Southeast are strictly carnivorous and eat only live animals, although some may occasionally eat plants such as berries or grapes. Small lizards prey primarily on insects, spiders, and other invertebrates. Larger ones, such as broad-headed skinks, will eat small vertebrates, including mammals, birds, and other lizards. All lizards have teeth, but the teeth are used more for grasping and positioning prey for swallowing

Southeastern lizards eat hundreds of thousands of insects each year.
A green anole in Lancaster County, South Carolina, eats a katydid.

Smaller lizards often fall prey to larger ones of the same or different species. An introduced black spinytail iguana in Florida eats an introduced rainbow whiptail.

than for chewing. Southeastern lizards typically use vision to locate prey, which they attack by biting and holding on, then swallowing the animal whole. Most are agile hunters that approach a potential meal stealthily and then lunge and grab it. Many lizards are categorized as dietary generalists because they are known or presumed to eat whatever prey is available and of appropriate size. Specialists, in contrast, maintain diets that focus on particular kinds of animals, which may include termites, ants, and even snails.

Herpetologists recognize two basic predation techniques. "Sit-and-wait" ambush hunters, which include the green anole and the geckos, remain motionless until an unsuspecting insect is within striking range, then attack. Active foragers, such as the southeastern skinks and the six-lined racerunner, cover large areas in search of prey, which they run down and seize. Western whiptail lizards, close relatives of the six-lined racerunner, commonly travel a half-mile or more each day while foraging. The two predation categories are not always distinctive, and some species may use both approaches, depending on the circumstances. Like other reptiles, lizards can go long periods—days, weeks, or even a few months during winter—without eating.

Reproduction

The reproductive cycle and mating behavior of lizards resemble those of other vertebrate animals, although a few exceptions are apparent. The basic reproductive pattern is for males to court females during the annual mating season, which begins in the spring in the Southeast. As is the case for many species of birds, the males of some lizard species are larger and more brightly colored than the females (especially during the mating season). Combat or threat displays between males are common, the bright red dewlap under the throat of the green anole being a classic example. All lizards have internal fertilization. The mating pair join their reproductive organs together while the male straddles the female. Male lizards, like snakes, have paired reproductive organs called hemipenes, which presumably ensure successful mating no matter which side of the female the male approaches.

From spring to midsummer, southeastern lizards lay eggs that hatch in late summer or early fall. Although most of the lizards found in the United States lay eggs, a few species, such as the pigmy short-horned lizard and the northern alligator lizard in the Northwest, give live birth to babies in late summer at the same time the young of egg-laying species are hatching.

Although this type of reproduction is unknown in native lizards of the Southeast, some species in the western United States and on other continents — as well as an exotic species introduced into the United States — are

Southeastern skinks, such as these mole skinks, mate aboveground in encounters that can sometimes appear like aggressive behavior.

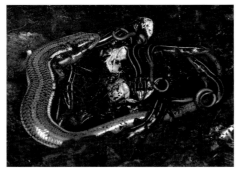

Some lizards show parental care. Female five-lined skinks of the Southeast guard their eggs until the young hatch.

Some lizard species, such as this Indo-Pacific gecko introduced into Florida, are parthenogenetic: only females occur in populations of the species.

Bright colors of adult males of many lizard species, such as this fence lizard, commonly differentiate them from adult females and juveniles of either sex.

parthenogenetic. That is, females reproduce asexually, producing eggs without mating with males. In fact, males are unknown in some species of western whiptail lizards. Parthenogenetic females produce only female offspring that are genetically identical with the mother and each other. The Indo-Pacific gecko introduced into Florida is an all-female species.

The leathery-shelled eggs produced by lizards are highly susceptible to predators, and females use a variety of strategies to protect them. Some take great care to conceal their eggs, burying them underground, in debris in tree holes, or in disturbed areas such as sawdust piles or piles of sand. Females of many of the skink species, including some in the Southeast, stay with their eggs to guard them against attacks by small predators or to move them around to prevent fungus from developing on the surface of eggs. Green anoles do not guard their eggs; instead, females lay only one egg at a time, each in a different place, reducing the chance of losing a whole clutch. It is the classic example of not putting all of your eggs in one basket. Eggs of native southeastern lizards usually hatch after a 4 to 7-week incubation period, with the length depending on the species, temperature, and moisture level.

Broken tails sometimes regenerate with a forked appearance, as seen in this green anole (left) and six-lined racerunner (right).

Defense

Like lizards elsewhere in the world, most southeastern lizards are small and fall prey to larger animals such as snakes, birds, and cats. Not unexpectedly, lizards have a vast array of defensive mechanisms to cope with the continual threat of becoming a meal for a larger predator. Southeastern lizards have several characteristic predator avoidance and defensive strategies. Most species are especially adept at the first and most effective step in avoiding predation: remaining undetected either by camouflage or by using behaviors that facilitate hiding underground or beneath ground litter or other cover. Once sighted by a predator, the next line of defense is to flee, usually toward a burrow, up or around a tree trunk, or into heavy vegetation. Racerunners can actually outrun or outmaneuver most large predators, including humans trying to catch them.

Most lizards will bite, and that may deter some predators, but tail breakage is by far the most common last-resort defense that is likely to be successful. The broken-off tail typically writhes on the ground and attracts the attention of the predator while the lizard itself makes a fast escape. The disproportionately long tail of the glass lizards and the bright blue or red tail of some skinks are presumably adaptations that make it likely that a predator will end up with the lizard's tail as a minor snack rather than making a meal of the entire lizard. Most lizards can regrow a broken tail, but the new tail seldom reaches the length of the original. In many species, a regrown tail is easily recognizable because it is a different color.

The Southeast Asian flying geckos or parachute geckos have webbed feet that can aid in escaping a tree-climbing predator. The lizard is able to leap from a limb and glide to the ground or another tree, buoyed by the membranous skin between the toes.

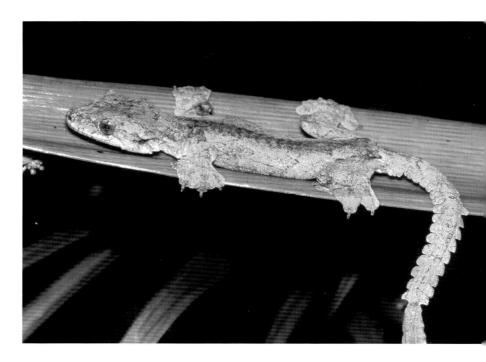

The brightly colored tail of some skinks has been suggested to be a warning sign to predators that the lizard itself may be poisonous to eat. Veterinary reports of cats becoming ill after eating blue-tailed skinks support this conjecture. Skinks also have a layer of osteoderms beneath the skin that creates a protective armor, an effective defense against many snakes and other animals whose teeth cannot penetrate or circumvent the latticework of bony scales.

Some unusual forms of defense have been reported among the exotic species now established in the Southeast. Horned lizards squirt blood from the eyes, green iguanas whip their tail, and green iguanas and spinytail iguanas swim or bottom-walk to escape predators. The brown basilisk, an exotic lizard species now found in southern Florida, literally runs to safety across the water surface. Lizards found elsewhere in the world have equally interesting defense behaviors. For instance, some geckos hold the tail above the back in mimicry of a venomous scorpion, and others lose the outer layer of their skin when grasped by a predator, allowing the lizard itself to escape. Open-mouthed threat displays, a common defensive mechanism of some Australian and South American lizards designed to ward off predators, are also not typical of southeastern species.

Predators

The fact that most of the lizards native to the Southeast have a breakable tail is an indication that they are a common prey item for many predators. Their most common natural predators are snakes, birds, and, in some instances, other lizards. Spiders, praying mantises, and centipedes will all prey on small lizards if given the opportunity. Imported fire ants, which have become widely established in the Southeast, may be a threat to the

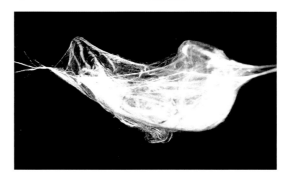

Juvenile lizards sometimes fall prey to web-building spiders.

eggs. Feral cats are major predators of lizards in the Southeast. Domestic housecats are among the most proficient of all lizard predators, not necessarily driven by hunger but simply because they have the leisure time to hunt for sport.

Several species of southeastern snakes eat lizards, although none eats lizards exclusively. Kingsnakes, coral snakes, and racers are all documented predators of lizards. Some snakes are especially

Domestic house cats are frequent predators on native and exotic lizards.

Lizards are food for many predators, including snakes. A scarlet kingsnake eats a little brown skink that lost its tail in the struggle.

effective as lizard predators because they can forage in shrubs or trees as well as beneath ground litter. Scarlet snakes eat reptile eggs and are likely predators of lizard eggs laid in sandy habitats. Broad-headed skinks and some other species of large lizards will prey on other lizards, even smaller individuals of their own species. The knight anole introduced into Florida is especially noted for eating smaller lizards, including other anoles.

Most of the birds that prey on lizards attack from above. Kestrels (sparrow hawks) have been recorded bringing six-lined racerunners to the nest to feed their young and have been observed eating anoles and fence lizards. Loggerhead shrikes are renowned for catching anoles or little brown skinks and then impaling the unlucky morsel on a thorn or the barbs of a barb-wire fence until they are ready to eat it. The greater roadrunner of the U.S. Southwest is a notorious predator of lizards, and the roadrunners that range into western Louisiana undoubtedly take their share of local lizards.

Lizards typically do not constitute a major portion of the diet of carnivorous mammals. Chasing a fast-moving lizard that disappears beneath the ground cover is not ordinarily a worthwhile pursuit for a large mammal such as a bobcat or coyote. Medium-sized species such as raccoons, skunks, and foxes, which are among the most notable predators of many other small vertebrates, probably also seldom bother to pursue lizards.

Most southeastern lizards do not congregate to hibernate as some snakes do and therefore are not vulnerable in concentrated numbers to predatory mammals searching for food during cold weather when reptiles are dormant. A dozen or more green anoles may congregate beneath tree bark or siding on abandoned barns or houses, however.

Habitats

Lizards live in most terrestrial habitats of the Southeast, including suburban, urban, and agricultural areas. The greatest diversity is found in natural habitats, particularly the mixed hardwood and pine forests that occur throughout the Southeast and the sandhill pine forests and other sandy vegetated areas of the Coastal Plain. The sand-burrowing

Sandy soil in the open habitat of a scrub oak forest is ideal for the six-lined racerunner.

mole skinks and most of the glass lizards are restricted to particular habitat types with the sandy soil that their underground lifestyle requires. Eastern fence lizards are more likely to be found in open areas at the margins of wooded habitat than in the interior, and individuals are frequently encountered around old barns, bridges, or fences. One population that has been the subject of a long-lasting research project lives next to a curved guardrail along a highway in a remote area of South Carolina. Other native species, such as green anoles, five-lined skinks, and little brown skinks, are relatively ubiquitous throughout much of their broad geographic range. The introduced exotic lizards that have taken up residence in Florida characteristically thrive in urban or otherwise heavily developed areas with dense vegetation.

Our native lizards are commonly found in the vicinity of streams, rivers, and lakes, but none are considered truly aquatic. Eastern glass lizards have been found in shallow water at the edge of Carolina bay wetlands, and southern coal skinks occasionally escape predators by entering water. No modern lizards live permanently in the ocean or brackish water habitats, although glass lizards are sometimes very common in sandy areas near coastal salt marshes.

The presence or absence of legs is an identifying trait for several native species of southeastern lizards, including the legless slender glass lizard.

The Florida reef gecko is the smallest lizard in the Southeast.

HOW TO IDENTIFY LIZARDS

Identifying native lizards in the Southeast is usually a relatively straightforward matter because of the limited number of species, although distinguishing between some of the skinks or some of the glass lizards can be problematic. Making a proper identification can be more difficult in some areas where introduced species are prevalent. Outside southern Florida, however, familiarity with a few particular characteristics should be sufficient to distinguish most species.

Legs

The absence of legs immediately differentiates the Florida wormlizard and the four species of glass lizards (*Ophisaurus*) from all other lizards in the Southeast, including introduced forms. Leg length in relation to body size can also be important. The sand skink, for example, has tiny, seemingly useless legs.

Body Size

Native southeastern lizards range in body size from little brown skinks not even 6 inches long from nose to tail to broad-headed skinks and glass lizards that may be twice that long and far more massive. Because they are among the easiest measurements to take on a lizard, body length and tail length are the typical size references, and we will use the two combined (total length) as the indicator of size. Whether or not an individual's tail has been broken should be taken into account when making identifications based on the total length. Because so many lizards lose their tails, the actual body length from snout to vent (cloacal opening) is often used as the standard measure of body size in ecological studies. The life stage, which may be difficult to determine because some lizards look like miniature adults from the time of hatching, must also be considered in identification. All lizards increase in length considerably from birth to adulthood, but juveniles of some species can resemble and be mistaken for adults of closely related species.

Body Shape

Some lizards can be easily characterized on the basis of body shape (e.g., true chameleons and horned lizards), but with the exception of the wormlike wormlizard and the snakelike glass lizards, most species native to the Southeast do not vary dramatically from one another in that regard.

The number and position of longitudinal stripes on the body are commonly used to identify many native species, such as this six-lined racerunner, as well as most of the skinks and glass lizards. The intensity of blue on the chin can be used to distinguish males from females of racerunners and fence lizards.

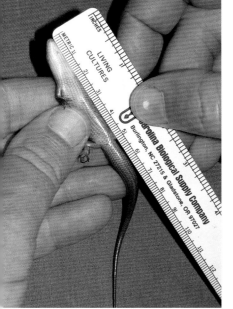

Body Markings

Some lizard species can be readily distinguished on the basis of their body markings. The presence of light-colored stripes down the back identifies some species as five-lined skinks or six-lined racerunners, and the position of dark stripes along the sides is important in differentiating among glass lizards. Markings such as spots, blotches, and bands characterize some kinds of lizards. Florida reef geckos have small dark spots, for instance, and some fence lizards have "chevrons" on the back.

A standard measurement by lizard research ecologists is the snout–vent length (snout to cloaca) in millimeters (left). The total length in inches (including the tail) is typically used to denote body length in field guides.

Color Patterns

Many southeastern lizards have colorful or unique body patterns that can be useful in accurately assigning an individual to a particular species or in eliminating others from consideration. The bright green body of the green anole is unmistakable, although the same individual may change to brown and be less distinctive. Like birds, males of many lizard species are more

A male green anole can display its red dewlap or throat fan even when its body is in the less conspicuous brown color state.

brightly colored than the females, especially during the breeding season. Males of some species do not change dramatically in color but instead develop a prenuptial blush that distinguishes them from females. The blue chin and belly of male fence lizards and racerunners, the dramatic red dewlap of green anoles, and the orange or red head of adult male five-lined and broad-headed skinks provide contrasts between the sexes as well as among species. A blue or red tail is also distinctive of certain species of skinks.

Skin Texture

The smooth, shiny scales of skinks and glass lizards are quite different from the rough scales of fence and scrub lizards and the granular scales of racerunners and anoles. Although not characteristic of any native southeastern species, lizards from other regions may have a beaded appearance (Gila monster), scales that are large plates (African plated lizards), or scales with spines (horned lizards).

Other Distinctive Characters

The absence of movable eyelids (reef gecko), wide belly scales (racerunners), and folds that run along the sides for the length of the body (glass lizards) are among the other traits that can be used to distinguish southeastern lizards.

Geographic Location

The geographic location in the Southeast where a lizard is found may be an instant clue to its identity.

A blue-tailed skink from southernmost Florida (including the Keys) will be a southeastern five-lined skink rather than one of the several other species with a blue tail. Knowing the region a species is from can often help eliminate particular species with which it might be confused morphologically. For example, any gecko found outside southern Florida will be an introduced exotic rather than the native reef gecko.

Habitat

A pair of broad-headed skinks (male on right) basks on a tree limb.

The actual habitat where a lizard is found may assist in its identification, although this information is often less helpful with southeastern lizards than with the other groups of reptiles. Knowing, however, that broad-headed skinks are often found in association with large live oaks, that little brown skinks are especially prevalent in forests covered with pine straw, and that glass lizards are most common in areas with sandy soil can all be useful clues for determining what species to expect where.

Time of Day and Year

Season is not generally a useful identification aid for southeastern lizards because most of the native species respond to, and become active in, fairly similar environmental conditions. While individual species may prefer slightly warmer or cooler temperatures, no strong seasonality differences are apparent among species. Likewise, lizards of the Southeast are active during daylight hours, with the exception of most geckos, which are so distinctive that they would be hard to confuse with the other lizard groups anyway.

The Texas horned lizard is not native to the Southeast, but local populations have become established in areas of North Carolina, South Carolina, and Florida.

The Chinese alligator is on the verge of extinction in the wild.

All About Crocodilians

CROCODILIAN BIODIVERSITY

Although there once were many more, only 23 species of crocodilians currently survive. All but 2 species are tropical (the American alligator and the Chinese alligator are temperate zone species), and nearly half—6 species of caimans and 4 species of crocodiles—occur in the New World tropics.

Crocodilians belong to three distinct families: Alligatoridae, Crocodilidae, and Gavialidae. Of the 8 species of the Alligatoridae, which includes the caimans, 7 live in the Americas, with the Chinese alligator again being an exception. The 14 species of the Crocodilidae are native to all major land masses in the tropics, including many islands. In addition to the several species of large to medium-sized crocodiles, the family also includes the dwarf crocodiles of tropical West Africa and the bizarre-looking, long-snouted tomistoma, or false gharial, of Indonesia and Malaysia. The Gavialidae comprises a single species that resembles the tomistoma: the highly aquatic, fish-eating Indian gharial, also called "gavial" due to an early misspelling in the historical literature.

Crocodilians do not vary appreciably in body size or shape. All have four limbs and a relatively large mouth with teeth that can deliver a bite that ranges from attention getting to deadly, depending on the animal's size.

When in the water, crocodilians such as this American alligator keep most of their body beneath the surface, making them inconspicuous to potential prey.

The most distinctive superficial difference among species is the shape of the snout, which ranges from the broad snout of the alligators and some of the caimans to the extremely long, slender snout of tomistomas and gharials. The largest species, the saltwater crocodile, sometimes reaches a length of 20 feet and ranges from Australia and Papua New Guinea through the Philippines and Indonesia to Sri Lanka and India. Cuvier's dwarf caiman of South America is the smallest crocodilian; adults do not reach a length of 6 feet.

Only two crocodilian species are native to the United States: the American alligator (Alligatoridae) and the American crocodile (Crocodilidae). Populations of a third species, the spectacled caiman (Alligatoridae), have become established in the aquatic systems around Homestead, Florida.

Despite the similarity in appearance among most species of crocodilians, the species found in the Southeast can be distinguished from each other by skull features, such as the narrower snout of the American crocodile (top) compared to the broader snouts of the American alligator (bottom) and introduced spectacled caiman (middle). The caiman has a bony ridge that runs across the snout in front of the eyes.

GENERAL BIOLOGY, ECOLOGY, AND BEHAVIOR OF CROCODILIANS

The many species of crocodilians that lived on Earth during the Triassic more than 200 million years ago did not differ much in appearance from the alligators, caimans, and crocodiles that we know today. Most had a long snout with big teeth; four well-developed legs with claws; a heavy, elongate body; and a powerful tail flattened for propelling the animal through water. A few of the extinct forms lived in the ocean and had paddlelike limbs, and some were terrestrial. Ancient forms, like modern ones, had osteoderms that formed an armor of bony scales beneath the thick skin of the body and tail. All living crocodilians have webbing on all four feet. Crocodilians continually shed and replace their teeth throughout their life.

Adult crocodilians are fairly nondescript in color, if not in size. They are either black or some shade of gray, brown, or olive on the back and sides, and white to yellowish on the belly. The young are more brightly colored, with yellow bands on black or black bands on gray or olive that fade as the animal grows larger. Crocodilians have well-developed upper and lower eyelids as well as a transparent third eyelid that moves laterally across the eye.

Crocodilian anatomy is typical of other reptiles in most respects. The heart is a major exception. In anatomically simplistic terms, crocodilians have a four-chambered heart whereas snakes, lizards, and turtles have a three-chambered heart. Crocodilians are well adapted for life in an aquatic environment and are able to see, hear, and smell underwater. Their pupils are elliptical, and their night vision is excellent. The nostrils are situated on the front of the snout as in other reptiles, but they are on the top rather than at the end, allowing them to breathe while they float in the water with the rest of the body submerged. Numerous sensory organs cover the body surface; some are sensitive to vibrations in the water. The function of two large glands that are readily apparent under the lower jaw has not been established.

Crocodilians can hear airborne sounds well and use them to communicate with one another. American alligators of both sexes and all sizes vocalize, the diversity of their vocal communications indicating the existence of social relationships. Hatchling alligators first vocalize while still in the nest. The sounds signal the mother alligator, who is usually nearby in the water, to crawl onto shore and open the nest, allowing the young to exit and enter the water. For a year or more after hatching, young alligators will emit a grunting sound when they perceive danger, to which the mother will respond, even leaving the water to attack a perceived predator. The bellowing of male alligators during the early spring alerts receptive females of the male's availability for mating—and perhaps of his suitability as a mate—and presumably also warns other males in the vicinity to stay away. Females sometimes bellow during the mating period, possibly in response to a calling male. Adults and juveniles continue to communicate throughout their lives with a wide variety of mostly low-pitched sounds that are audible through the air or underwater and are thought to convey different messages. Crocodilian social behavior, while instinctive rather than learned, is highly complex and not thoroughly understood by behavioral scientists, offering many opportunities for study.

Did you know?

Despite claims of alligators that are more than a century old, no scientifically documented record exists of an alligator more than 75 years old.

Despite having formidable claws on all feet, alligators do not normally use them defensively to scratch.

Activity and Locomotion

Crocodilians that live in tropical and subtropical climates are active throughout the year. The American alligator, in contrast, experiences seasonal changes in temperature and is inactive from November until March in the most northern and inland parts of the geographic range outside southern Florida. Alligators spend most of the winter underwater in a torpid state, but they must take in air on a regular basis and therefore rise to the surface periodically to breathe. Even in the coldest parts of their range, some individuals will emerge from the water to bask on sunny days if the air temperature is warm enough.

Crocodilians are typically active both day and night. American alligators and American crocodiles commonly bask on shore or swim during the day and stay in the water and hunt at night. If the opportunity presents itself, however, both species will capture prey in the daytime.

A swimming alligator uses its tail as a rudder to propel and guide itself.

Crocodilians swim by using the long, vertically compressed tail as both a rudder and a propeller. Although the feet are webbed, the legs are used minimally or not at all for propulsion; the limbs trail alongside the body when the animal is moving through the water. All crocodilians are able to walk overland, sometimes traveling substantial distances from one body of water to another. They can run across the ground in speedy bursts for short distances, but none of today's crocodilians would be able to outrun a human inspired by a modern-day dinosaur in pursuit.

An adult alligator moves overland by "high walking" from one wetland to another.

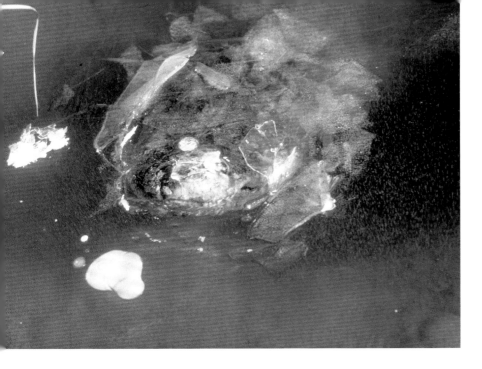

An adult alligator at the Savannah River Ecology Laboratory in South Carolina has positioned itself with its snout above the ice of a lake where it may remain motionless for days during winter.

Temperature Biology

Crocodilians are ectotherms, and they respond to temperature changes as lizards do (see page 26). American alligators can survive for long periods at water temperatures below 39°F (4°C), as evidenced by observations of adults with their snouts frozen in surface ice during unusually cold winters in North and South Carolina. Crocodilians benefit from being ectotherms in the same ways that lizards do (see page 27).

Food and Feeding

Crocodilians are strictly carnivorous. They catch live prey—especially fish, turtles, birds, and mammals—but also feed readily on carrion. Prey is captured both in the water and on land at the water's edge. Crocodilians are stealth hunters that approach their prey inconspicuously on the water's surface and then submerge before attacking. Like other reptiles, crocodilians do not chew their food but instead swallow it whole. If a prey item, such as a deer, is too large to consume in one gulp, an alligator will cache the carcass under the stream or lake bank and return to consume it when the flesh has begun to decay and is more pliable. Female alligators have been observed providing what may be a form of parental care to their young, which will often gather around her mouth while she is feeding and eat scraps that fall from her mouth.

Among the crocodilians are a few species that actively stalk and attack humans. Extensive unequivocal documentation exists of large saltwater

A 12- to 14-foot alligator carries a deer across the U.S. Fish and Wildlife Service's Harris Neck National Wildlife Refuge in Georgia. Adult alligators commonly prey on full-grown deer that come too close to the water's edge or swim across open water.

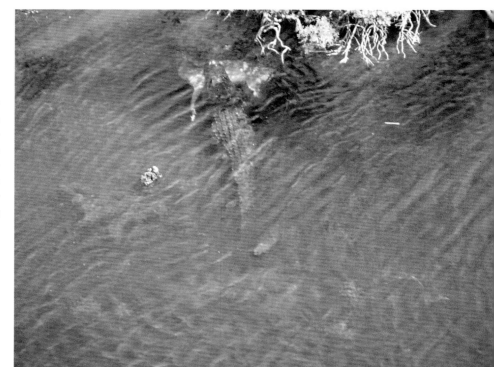

A saltwater crocodile leaps from the water on the Adelaide River near Darwin, Australia, to snare a chunk of pork dangled from a pole over the edge of a boat.

crocodiles of Australia and Indonesia eating both adults and children. People have been attacked while swimming, while on the shore, and while in small boats. American alligators do not normally attack humans unless they are provoked or are defending their eggs or young. Nevertheless, it is sensible to be cautious around alligators, crocodiles, and caimans, as credible reports have been made of several species, rarely including American alligators, attacking humans as prey. American alligators consider dogs suitable prey at any time, though. They will grab dogs that are swimming or playing in the water and will even come out of the water after them, frequently while the dog is quenching its thirst for the last time.

Reproduction

All modern crocodilians lay eggs, and all the species studied nest in prescribed seasons. Females of some species take more than 15 years to reach maturity, and most take at least 10. Males of some species, such as American alligators, become much larger than females and engage in combat during the breeding period. Elaborate courtship behavior prior to mating has been described in alligators and other crocodilians.

Reproductive cycles of crocodilians in tropical areas respond to annual wet-dry cycles, with some species, including the American crocodile, laying eggs that incubate in the moist, warm nest during dry weather and hatch at the onset of the wet season. Other species lay eggs during the wet season. American alligators engage in courtship activities in early spring and lay eggs in the early summer. The average number of eggs varies from as few as a dozen to four dozen, the higher numbers not surprisingly being laid by the larger species and individuals. Females of most species lay their eggs in mound nests that they build of soil and vegetation, but those of some species, including the American crocodile, dig nest holes in which they deposit and then cover the eggs the way most turtles do. Several species are known to have an incubation period of 2–3 months, but the

These baby alligators bask safely on their mother's back. Alligators have been known to protect their young for more than a year after hatching.

period may be shorter or longer under different temperature and moisture conditions. The females of several crocodilian species, including American alligators, open the nest when the babies hatch and gently carry them to the water in their mouth.

The sex of developing crocodilians is not determined by sex chromosomes, as is the case for mammals. Instead, crocodilians exhibit "environmental sex determination," a phenomenon characteristic of many turtles, in which the nest incubation temperature influences the sex of the hatchlings. For example, in some species of crocodilians, including American alligators, a nest temperature around 90–91°F (32–33°C) produces males, whereas temperatures lower than about 88°F (31°C) or above about 93°F (34°C) produce females. (The precise temperatures vary among species and are unknown for most.)

Crocodilians around the world have been reported to guard their nests against predators and to protect the young for days or months after hatching. Parental care has not been fully studied in most species of crocodilians, but alligators have been known to protect their young for more than a year.

The female American alligator lays
eggs inside a large aboveground
nest she builds near a wetland (top).
Baby alligators hatch from the eggs
(bottom left and right) in late
summer or early fall.

Humans are one of the
greatest threats faced by
adult crocodilians throughout the world. Even the
largest individuals can be
snared and captured.

Predators and Defense

Under natural conditions, crocodilians are most vulnerable during the egg stage. Terrestrial mammals can dig the eggs out of the nest, and ants can be a problem in some situations. Although a female can defend her nest from some predators, she can do little to stop fire ant predation. Nor can she prevent high water from drowning the eggs or extreme drought or heat from desiccating or overheating them.

The aquatic young are vulnerable to large fish, turtles, snakes, wading birds, raptors, and aquatic carnivorous mammals. Both American alligators and American crocodiles have been known to eat smaller individuals of their respective species. Contrasting body coloration provides camouflage for the young when they are in the aquatic habitat, and most juveniles receive some protection from their mother. When confronted by enemies, the young of some species will swim into under-the-bank dens dug by their mother.

Crocodilians beyond the early juvenile years have few predators. Large constrictors (anacondas in South America and introduced Burmese pythons in the Florida Everglades) have been documented eating 6-foot-long caimans and alligators, respectively. Larger individuals of any species of crocodilian will presumably overpower and eat smaller individuals of the same or different species if given an opportunity. Man, however, is probably the greatest threat faced by most adult crocodilians.

Crocodilians of any age or size will attempt to escape into deeper water or beneath a bank when threatened, even though most quickly reach a size that places them beyond any threat of becoming prey for another animal. Few predators can penetrate the tough outer covering of skin underlain by osteoderms on the back, sides, and tail. Osteoderms also protect the belly of some species. All crocodilians can bite, and most adults are equipped with large teeth and have crushing jaws that can subdue almost any other animal that might attack them. The muscular tail of the largest species can easily break an adult human's leg and would be a deterrent to any animal.

A loudly hissing mother alligator charges open-mouthed from the water to protect recently hatched babies that have retreated beneath the bank.

Baby crocodilians, like this young alligator, have a full set of teeth and are prepared to bite an attacker, but their jaw strength is usually not enough to break the skin until they approach about a foot in length.

An adult American alligator basks alongside a coastal waterway.

Habitats

Crocodilians are less diverse than lizards in terms of habitats occupied. All are associated with freshwater aquatic areas. Some species will enter brackish or saltwater habitats, and some individual saltwater crocodiles travel long distances in the ocean. The American crocodile can persist in estuaries, other brackish water systems, and even in waters of high salinity. American alligators prefer freshwater but will enter brackish water or salt water.

Alligators are found in a wide variety of aquatic habitats throughout their range in the Southeast.

A startled alligator (above) dives into a brackish water habitat on Capers Island, South Carolina. Alligators usually remain in freshwater but in coastal areas will enter brackish or salt water to feed.

American alligators and other crocodilians will bask aquatically while remaining alert for an opportunity to capture surface-swimming prey such as snakes, turtles, or waterfowl.

Species Accounts

ORGANIZATION AND ORDER OF SPECIES ACCOUNTS

The species accounts that follow are designed to familiarize the reader with the native and introduced species of lizards and crocodilians of the Southeast. The two major groups, lizards and crocodilians, are presented in taxonomic order by family, with the native species (six families of lizards, two families of crocodilians) preceding the introduced species that have become established (eight families of lizards, one family of crocodilians). Within each family, the species accounts are ordered taxonomically by genus and alphabetically by species. Numerous changes in the taxonomy of lizard genera were made just before and after the turn of the twenty-first century. We use the genus name, whether the original or proposed, that we consider likely to become accepted by lizard biologists. We mention alternate names, including those commonly used for some species in the early 2000s, in the discussion of taxonomic issues. See the indexes of scientific and common names for the locations of particular species accounts.

The species accounts include information on appearance, distribution, ecology, and behavior. Maps showing the distribution accompany each native species account. The larger map shows the species' range within the Southeast and the smaller map illustrates the entire range within the United

The green anole may be the most commonly seen native lizard in the Southeast because of its ability to thrive in urban and suburban areas.

States. The maps represent distribution patterns based on county records that have been determined by herpetologists within each of the southeastern states. The contemporary distribution patterns of some species may not match historical records because many native species have disappeared from large regions of the Southeast, victims of increasing human development and landscape modification. The geographic range for some native lizards will thus include areas where the species no longer occurs. In contrast, the currently stated range for many exotic lizards—and some native species—may increase as populations spread through human transport.

The accounts for the introduced species are briefer than those for the native species because information on most of them did not begin to emerge until the 1990s and early 2000s, and is in almost all cases incomplete. Quite often

BASIC FEATURES OF THE SPECIES ACCOUNTS

Quick identification guide

Adult male eastern glass lizards are often greenish in color.

How do you identify an eastern glass lizard?

BODY PATTERN AND COLOR
brown or greenish above with greenish white speckles on the sides; longitudinal dark stripes usually present on sides, but only above the lateral fold

DISTINCTIVE CHARACTERS
old adults are often greenish with yellow belly

SIZE

● ADULT
● HATCHLING

Eastern Glass Lizard
Ophisaurus ventralis

FAMILY Anguidae

DESCRIPTION Eastern glass lizards have no dark stripe down the center of the back and may lack dark stripes altogether. Adult males often are greenish overall with white dots giving them a vivid speckled appearance; younger individuals and females may have dark longitudinal side stripes, but not below the lateral fold. White dash marks are evident on the sides of the neck of juveniles and some adults. The belly is yellow to cream. Many specimens have a distinctive regenerated tail tip that is plain brown. More than 97 body scales are present along the lateral fold.

VARIATION AND TAXONOMIC ISSUES No subspecies have been described and no geographic variation in color or body morphology has been noted.

WHAT DO THE HATCHLINGS LOOK LIKE? Hatchlings are beige to khaki colored, some with a greenish brown tint, and have a longitudinal dark stripe down each side.

CONFUSING SPECIES The lack of a dark stripe down the center of the back distinguishes eastern glass lizards from slender glass lizards and mimic glass lizards. The island glass lizard has one distinct stripe above the lateral fold and is immaculate below it, while the eastern glass lizard may have

86 · *Eastern Glass Lizard*

their ecology is not well known even in their native countries, and the most complete information available is based on the experiences of people in the pet trade. The shifts in ecological and behavioral traits that can be expected of species as they are introduced into new environments on a new continent are unpredictable and are still being documented. It is a vast experiment whose results will not be known for years. Not only is the complete biology of these translocated species poorly known, it is questionable whether they will even survive indefinitely in their new homes.

The majority of introduced species have localized distributions in southern Florida that are identified in the account by county references. Maps of recently introduced species could be misleading because the biological process of becoming a successfully introduced exotic species is a dynamic one.

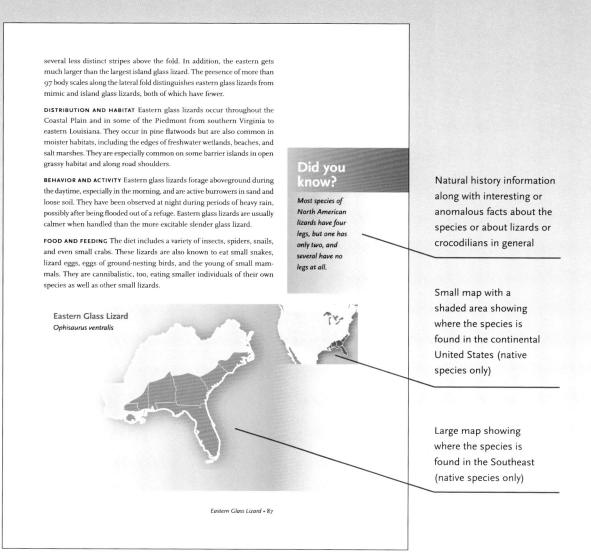

several less distinct stripes above the fold. In addition, the eastern gets much larger than the largest island glass lizard. The presence of more than 97 body scales along the lateral fold distinguishes eastern glass lizards from mimic and island glass lizards, both of which have fewer.

DISTRIBUTION AND HABITAT Eastern glass lizards occur throughout the Coastal Plain and in some of the Piedmont from southern Virginia to eastern Louisiana. They occur in pine flatwoods but are also common in moister habitats, including the edges of freshwater wetlands, beaches, and salt marshes. They are especially common on some barrier islands in open grassy habitat and along road shoulders.

BEHAVIOR AND ACTIVITY Eastern glass lizards forage aboveground during the daytime, especially in the morning, and are active burrowers in sand and loose soil. They have been observed at night during periods of heavy rain, possibly after being flooded out of a refuge. Eastern glass lizards are usually calmer when handled than the more excitable slender glass lizard.

FOOD AND FEEDING The diet includes a variety of insects, spiders, snails, and even small crabs. These lizards are also known to eat small snakes, lizard eggs, eggs of ground-nesting birds, and the young of small mammals. They are cannibalistic, too, eating smaller individuals of their own species as well as other small lizards.

Did you know?

Most species of North American lizards have four legs, but one has only two, and several have no legs at all.

Natural history information along with interesting or anomalous facts about the species or about lizards or crocodilians in general

Eastern Glass Lizard
Ophisaurus ventralis

Small map with a shaded area showing where the species is found in the continental United States (native species only)

Large map showing where the species is found in the Southeast (native species only)

NATIVE LIZARDS

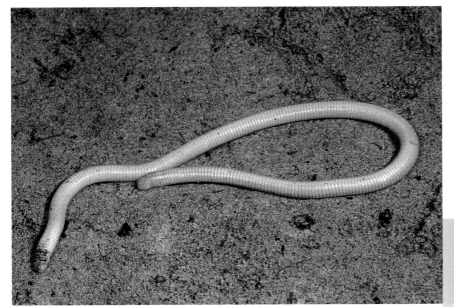

Florida wormlizards can be positively identified by a flattened tail tip and the presence of scales that encircle the body.

Florida Wormlizard

Rhineura floridana

FAMILY Amphisbaenidae

DESCRIPTION Florida wormlizards look more like pale pink earthworms than like any other true lizard found in the Southeast. A close look will reveal scales that encircle the body and are larger in the head region. Small grooves between the body scales create a ringed appearance. The absence of legs, ear openings, and (usually) visible eyes makes these unusual little animals distinctive. The flat, rough tail tip is an effective burrow plug if a predator is in pursuit.

VARIATION AND TAXONOMIC ISSUES Although the wormlizards found in south-central Florida differ both morphologically and genetically from those found farther north, only one species is recognized within the relatively small geographic range. Some lizard biologists place *Rhineura floridana* in a separate family (Rhineuridae), and some place it in a subfamily (Rhineurinae) within the family Amphisbaenidae. Most agree that the wormlizards are a distinctive group, although modern genetics studies have placed them evolutionarily with other lizards. The Florida wormlizard should arguably be called the southeastern wormlizard because the species is known from southern Georgia in addition to Florida.

BODY PATTERN AND COLOR
pinkish gray

DISTINCTIVE CHARACTERS
looks superficially like an earthworm and lacks limbs and ear openings; eyes absent or greatly reduced in size

SIZE

10"   4"

● ADULT
● HATCHLING

WHAT DO THE HATCHLINGS LOOK LIKE? Hatchlings look like small, light purplish earthworms and have tiny eyes that are not visible in adults.

CONFUSING SPECIES No other lizard in the United States is likely to be confused with this species.

DISTRIBUTION AND HABITAT Florida wormlizards are found in peninsular Florida above Lake Okeechobee and as far north as southern Georgia. The typical habitat is mixed hardwood or pine forests with little ground cover and deep sandy soils or sandy loam suitable for burrowing.

BEHAVIOR AND ACTIVITY The Florida wormlizard is completely fossorial, living its entire life in small tunnels it digs in the soil. Occasional specimens are discovered aboveground as a consequence of soil disturbance or flooding. Specimens are often found under deeply buried cover objects such as logs, deep leaf litter, or railroad ties.

FOOD AND FEEDING Florida wormlizards eat small invertebrates that they encounter underground or beneath leaf litter, including earthworms, spiders, and termites.

REPRODUCTION Few studies have been done on mating behavior or most other aspects of reproduction in this infrequently seen species. The females

Did you know?

When attacked by a predator, the largest neotropical wormlizard, known as the Brazilian "two-headed snake," simultaneously lifts its head and tail, giving it the appearance of having two heads.

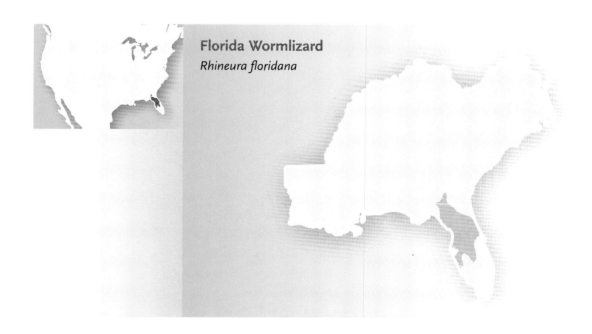

Florida Wormlizard
Rhineura floridana

Florida wormlizards lack obvious ear openings and have only the vestiges of eyes.

characteristically lay no more than three tiny eggs, more often only two or even one. Eggs are laid in the summer and have been found in sandy soils more than a foot below the surface. The young hatch in early fall.

PREDATORS AND DEFENSE Known predators include mockingbirds and loggerhead shrikes, but presumably coral snakes and other species that specialize on snakes and lizards would eat these soft-bodied creatures when they encounter them. In contrast to most lizards, wormlizards have a comparatively short tail that does not break when grabbed by a predator.

CONSERVATION ISSUES Too little is known of the ecology and behavior to recommend conservation measures beyond the protection and appropriate management of the sandy upland habitats that appear to be important to the species' existence.

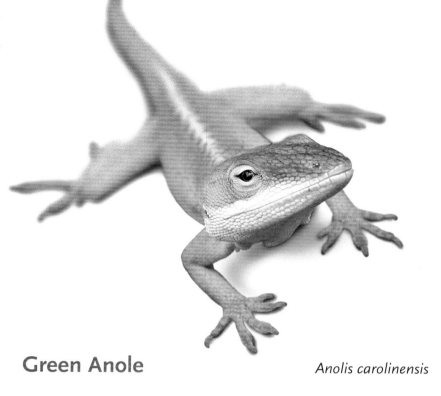

Both city dwellers and backwoods naturalists might stumble upon the color-changing green anole.

BODY PATTERN AND COLOR
green, brown, or gray with white belly

DISTINCTIVE CHARACTERS
dewlap of adult males usually red; ability to change body color from green to dark brown

SIZE

7"   1"

● ADULT
● HATCHLING

Green Anole

Anolis carolinensis

FAMILY Iguanidae

DESCRIPTION These medium-sized, long-snouted lizards are seen more commonly than most other species, often in urban and suburban areas. Juveniles and adults can change color from bright green to dark brown or dull grayish brown. The color change is a biologically complex response to environmental temperatures and social interactions. During the transition between green and brown, individuals occasionally exhibit a blotchy mix of the two colors, and females have a light stripe running along the length of the spine. The underside of the chin and the belly is white. The long toes have toe pads and distinct claws. The complete tail is typically more than 60 percent of the total body length. Males are particularly conspicuous during the breeding season when they assume a territorial posture to ward off other males. The most obvious territorial display is the male's solid-colored dewlap, which in most populations is bright red or reddish pink.

Hatchlings look like miniature adults.

Green Anole
Anolis carolinensis

VARIATION AND TAXONOMIC ISSUES Two subspecies have been recognized: the northern green anole (*A. c. carolinensis*) and the southern green anole (*A. c. seminolus*). The dewlap of the latter is much paler than the bright red of the more common subspecies. The southern variety is restricted to western counties above the Everglades in southern Florida.

WHAT DO THE HATCHLINGS LOOK LIKE? Hatchlings look like miniature adults and can also change their body color.

CONFUSING SPECIES Outside peninsular Florida and the expanding geographic range of brown anoles, green anoles are unlikely to be confused with any other lizard. Brown anoles do not turn green, and their dewlap

The green anole can turn brown during cool weather or under other conditions that are not fully understood, whether the background is brown or green.

A green anole male
displays his red dewlap.

is orange, yellowish, or burnt umber. The Cuban green anole in southern Florida is almost identical with the green anole in appearance, but each of the several other species of exotic anoles in southern Florida has a different and unique coloration. The color of the dewlap is a reliable identification character.

DISTRIBUTION AND HABITAT Green anoles occur throughout the coastal southeastern states from South Carolina to Louisiana, except for mountainous portions of northern Georgia. They are also found in southern Tennessee and the Coastal Plain of North Carolina, but the range does not extend into Virginia. They are associated with green vegetation, including vines, shrubs, and small trees. Individuals will climb high up into large trees and in urban or other developed areas are commonly seen on walls or around houses and buildings with ivy or other climbing vegetation. In forested habitats, they are also in areas with low-lying vegetation and can be common on plants around the margins of wetlands or other open areas. Green anoles have been introduced into Hawaii and have become established in several locations.

BEHAVIOR AND ACTIVITY These superb climbers may be active in any month of the year if temperatures are mild or if they can warm themselves sufficiently in a protected patch of sunlight. When active, they can often be seen in the open on the leaves of heavily vegetated habitat or on limbs, leaves, or other structure. Their foraging behavior is fascinating. Hunters stealthily move toward an unsuspecting insect, then lunge and seize it. Males are territorial and will display their dewlap and bob their head vigorously at an approaching lizard of either sex or size. Some will even threaten a person who is not close enough to reach them. Green anoles frequently congregate for hibernation beneath the bark of dead trees or under rocks.

FOOD AND FEEDING The diet includes a wide variety of insects, flying ones as well as caterpillars, spiders, scorpions, and other small invertebrates. Green anoles are reported to be attracted to carcasses of lizards of other species, possibly as scavengers or to eat the maggots that may be present. In addition to eating animals, green anoles lick nectar and tree sap, eat pollen and flower petals, and have even been observed lapping up the liquid at hummingbird feeders.

Did you know?

People in the Southeast often call green anoles "chameleons" because of the lizards' ability to change color.

Green anole males spar to defend their territories on Palmetto Bluff in South Carolina.

Mating occurs throughout the month of April and lasts through late summer.

REPRODUCTION Courtship and mating begin by April in all parts of the Southeast—even earlier in the southern part of the range on warm days in winter—and continue throughout the spring and summer. The males display continually as they defend territories and court resident females. Females dig a small depression in loose substrate in which to lay a solitary egg and then cover the nest. They continue laying eggs at approximately 2-week intervals from spring through summer, with location and summer weather conditions introducing notable variability. Communal nesting sites containing several eggs of the same or different females have been found. Most were out of sight beneath decaying vegetation or wood, under rocks, or even in the soil of potted plants. Baby anoles appear throughout the latter part of summer and fall.

PREDATORS AND DEFENSE The known predators include common kingsnakes, larger lizards, loggerhead shrikes and other birds, and wolf spiders. Presumably any predator large enough to capture a rather defenseless lizard could make a meal of these common lizards. Domestic cats and dogs commonly kill anoles as well.

CONSERVATION ISSUES Green anoles seem to fare well in just about any habitat as long as green vegetation is available. The most obvious threat is in Florida, where introduced exotic species of lizards, especially brown anoles, prey on and compete with them for food and habitat.

Eastern fence lizards are found in all southeastern states and are a commonly seen species in both suburban and rural areas along fence lines and around wood and brush piles.

How do you identify a fence lizard?

BODY PATTERN AND COLOR
generally gray or brownish

DISTINCTIVE CHARACTERS
blue coloration on chin and belly (pale in females, bright in males)

SIZE

7"   2"

● ADULT
● HATCHLING

Eastern Fence Lizard   *Sceloporus undulatus*

FAMILY Iguanidae

DESCRIPTION These common, medium-sized, gray- or brown-patterned spiny lizards are frequently spotted scurrying up trees and fence posts. Once they stop moving they can be hard to detect due to the natural camouflage endowed by the dark, wavy crossbars on their body. The females and young exhibit more contrast and are darker than the more brownish males, which also have bright light blue or greenish patches underneath. Some females have small patches of pale blue or green on the throat. Fence lizards have heavily keeled scales, each with a spine, that give their skin a rough appearance and feel.

The sex of adult fence lizards is easily determined because the undersides are bright blue in males, as shown here, and mostly plain gray or whitish with little blue in females.

VARIATION AND TAXONOMIC ISSUES Two of the 10 subspecies that have been described are found in the Southeast: the northern fence lizard (*S. undulatus undulatus*) and the southern fence lizard (*S. u. hyacinthinus*). Some authorities believe that the northern and southern subspecies are not biologically distinct, and some of the subspecies have been recognized as more closely related to other species of *Sceloporus* than to *S. undulatus*. Further work is needed to sort out their true relationships.

WHAT DO THE HATCHLINGS LOOK LIKE? Hatchlings resemble adult females, but with a darker and less distinct pattern.

CONFUSING SPECIES Within the Southeast, the only lizard that might be confused with this species is the Florida scrub lizard, which is smaller and has a dark brown stripe down each side.

DISTRIBUTION AND HABITAT Eastern fence lizards are found throughout the Southeast except in a strip of eastern Louisiana, all of coastal Louisiana, and southern Florida. In most of its range, this is the only species of spiny lizard. They are often abundant in dry, rather open woodlands and can be especially common on rocky hillsides and bluffs. While more common in pine forests, eastern fence lizards are also occasionally found in hardwood forests and in cutover areas where downed trees, stumps, and brush piles provide bask-

Hatchlings resemble adult females.

This male fence lizard has lost his tail.

Eastern Fence Lizard
Sceloporus undulatus

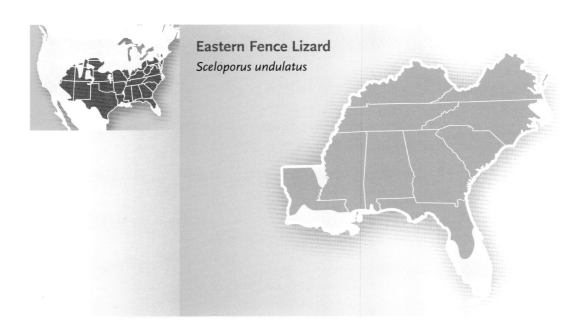

Fence lizards in the eastern portion of the geographic range (left) differ in appearance from those in the western portion (right).

ing sites and cover. These lizards tolerate some human disturbance and may be found around old farm buildings and lumber piles.

BEHAVIOR AND ACTIVITY Males are territorial and display their bright blue throat and belly by bobbing their head. They are excellent climbers. When disturbed they will dash to the nearest tree, often to the opposite side of the trunk, and into the upper branches, where they are soon out of the reach of predators. Mating occurs in spring soon after the adults emerge from their winter inactivity.

Fence lizards are common in pine forests.

FOOD AND FEEDING The diet includes beetles, ants, spiders, grasshoppers, millipedes, snails, and other invertebrates, and sometimes small lizards such as little brown skinks as well. Eastern fence lizards are "sit-and-wait" predators that sit motionless until prey comes within striking range.

REPRODUCTION Eastern fence lizards have been observed mating in the spring and early to late summer. The females bury 6–15 eggs in a shallow nest or burrow in loose sand, soil, or old sawdust piles and may lay multiple clutches within a year. No further maternal involvement occurs after the eggs are laid. The eggs incubate for 60–80 days, and hatchlings begin emerging from nests in midsummer and continue through early fall.

PREDATORS AND DEFENSE Known predators include a variety of snakes, birds, dogs, cats, and other predatory mammals. Low-intensity forest fires are a common threat to fence lizards and other species that live in habitats where natural fires are frequent. Survival responses to an approaching forest fire include climbing rapidly to the top of the canopy and burrowing into soft soil.

The mixed dark and light pattern of fence lizards helps camouflage them in their natural habitat.

CONSERVATION ISSUES Eastern fence lizards survive well in some disturbed habitats such as cutover or thinned woodlands and around old farms and outbuildings. Their populations are doubtless affected by the same factors that affect most other upland reptile species: habitat fragmentation and loss, and predation by house pets. They are, however, still fairly common throughout their historical range.

The brown longitudinal stripe on the body differentiates Florida scrub lizards (male, left; female, right) from eastern fence lizards.

How do you identify a Florida scrub lizard?

BODY PATTERN AND COLOR generally gray or brownish with a lengthwise brown stripe on each side

DISTINCTIVE CHARACTERS keeled scales and brown stripe along the upper sides of the body

SIZE

5"   1.5"

● ADULT
● HATCHLING

Florida Scrub Lizard   *Sceloporus woodi*

FAMILY Iguanidae

DESCRIPTION These small to medium-sized, gray or brown lizards have a distinctive broad brown stripe on each side of the body from behind the ear opening all the way to the base of the tail. Both sexes are mostly grayish on the belly except for a small patch of blue on the outer edges and on the throat. The blue patch is more pronounced in males and has a black border. Females have dark, irregular bands across the back. Like other fence lizards, Florida scrub lizards have keeled scales that give them a rough appearance and feel.

VARIATION AND TAXONOMIC ISSUES No subspecies are recognized. In areas where their ranges and habitats overlap, Florida scrub lizards may breed with eastern fence lizards and produce hybrids that are intermediate in appearance.

WHAT DO THE HATCHLINGS LOOK LIKE? Hatchlings look like tiny adult females, but their dark stripes are not as evident.

CONFUSING SPECIES The Florida scrub lizard might be confused with the eastern fence lizard in the central Florida peninsula where both occur, but fence lizards get larger, and if side stripes are present, they are not brown.

Florida Scrub Lizard
Sceloporus woodi

Both male and female Florida scrub lizards have blue patches on the outer edges of their bellies and on their throats.

DISTRIBUTION AND HABITAT Florida scrub lizards are found in some coastal areas of southern Florida and on sand ridges in the central part of the peninsula. They live almost exclusively in sand habitats, including coastal sand dune areas and sand pine–scrub oak forests, and are particularly associated with habitat having rosemary, a large shrublike plant native to sandy terrain in Florida that has historically been maintained under natural conditions by periodic fires.

BEHAVIOR AND ACTIVITY These lizards are active during the day, when they can be seen moving about on the ground. They occasionally sit on the trunks of small trees, primarily pines, either basking or waiting for prey to approach. When threatened, they escape by running across the sand and hiding beneath a bush or running up a tree. An adult male will confront another lizard or even a person by bobbing its head.

FOOD AND FEEDING Florida scrub lizards eat insects—notably ants, beetles, crickets, and grasshoppers—and spiders.

REPRODUCTION Mating may begin as early as February. Head bobbing is particularly prevalent during the spring mating season as males square off against one another. Females begin laying eggs in April, digging nests in the sand where they lay an average of four (two to eight) eggs before covering the hole. They typically lay two or three clutches of eggs in a year and possibly as many as five, occasionally as late as August. The eggs incubate underground, taking up to 10 weeks if ground temperatures are cool before the hatchlings begin to emerge in June, continuing into November.

PREDATORS AND DEFENSE Known predators include black racers and eastern coachwhip snakes that live in the same sandy habitats. Other lizard-eating snakes and predatory birds are likely threats to these small lizards as well. The primary defense of the Florida scrub lizard is to flee. The tail will break off if grabbed by a predator.

CONSERVATION ISSUES Florida scrub lizards have a small geographic range and occupy a habitat type that is rapidly vanishing in the southern half of Florida. Commercial development and agricultural conversion have already extirpated many populations. Protection of scrub habitats from environmentally insensitive urban or agricultural development offers the best assurance for the survival of this species in the wild.

Florida scrub lizards are active during the day—on the ground or on vegetation where they bask or wait for prey to come within range.

The six bright yellow stripes on the back and sides of the dark body distinguish six-lined racerunners from other southeastern species of lizards.

How do you identify a six-lined racerunner?

BODY PATTERN AND COLOR
dark back with light-colored stripes and white or blue belly

DISTINCTIVE CHARACTERS
six distinct yellow stripes on sides running length of the dark body and a faded or obvious center stripe down the back; males have a blue chin and belly

SIZE

8" 3"

● ADULT
● HATCHLING

Six-lined Racerunner   *Aspidoscelis sexlineatus*

FAMILY Teiidae

DESCRIPTION Six-lined racerunners are medium-sized lizards. The stream-lined body is dark gray to nearly black and has six conspicuous yellow or whitish stripes running from the eye to the tail. The brownish tail usually has one or two light stripes on the sides. The scales have a granular texture above and are in eight parallel rows of rectangular scales down the belly. The belly is white in females and juveniles and takes on a bluish hue in adult males.

VARIATION AND TAXONOMIC ISSUES Taxonomists have replaced the genus name *Cnemidophorus* used for North American species—including the six-lined racerunner—with the name *Aspidoscelis* on the basis of genetic analyses. Three subspecies have been described: the eastern six-lined race-runner, *A. s. sexlineatus*; the Texas yellow-headed racerunner, *A. s. stephensae*; and the prairie racerunner, *A. s. viridus*. The body of the prairie racerunner, the western subspecies, has a green tint on the front. The eastern subspecies (*A. s. sexlineatus*) is found in all southeastern states, but in Louisiana it intergrades with the western form, *A. s. viridus*.

Common names used locally for this species include sand streak, sand

runner, and field streak. All of the *Aspidoscelis* species in the U.S. Southwest are referred to as whiptail lizards.

WHAT DO THE HATCHLINGS LOOK LIKE? Hatchlings look like the adults but with even brighter stripes and a pale blue tail.

CONFUSING SPECIES The only lizards that might be confused with this species in the Southeast are the broad-headed and five-lined skinks, which have only five stripes down the body and smooth, shiny scales.

The pale blue color visible on the chin and belly of this male six-lined racerunner will become brighter during the spring mating season.

DISTRIBUTION AND HABITAT The six-lined racerunner is found thoughout most of the Southeast but is absent from parts of Virginia, North Carolina, Tennessee, Kentucky, and Louisiana. The typical habitat is open and dry and includes coastal sand dunes, longleaf pine–turkey oak sandhills, and some agricultural fields and forest clearcuts. Animals associated with heavily forested areas are most likely to be along open, sunny margins in sandy or rocky terrain.

BEHAVIOR AND ACTIVITY Six-lined racerunners are usually encountered when they are basking or foraging in open, sparsely vegetated areas or while zipping straight across paved or dirt country roads. If pursued, they escape with a rapid burst of speed to a refuge beneath vegetation, a rock, or a log, or into a burrow. They typically become active when temperatures rise during the morning and are likely to be seen at any time during a warm or hot day, even when their body temperature rises above 100°F (38°C). They hibernate earlier and emerge later than most other southeastern lizards, spending the winter in burrows they dig in sand or soft soil, often in the sunny south side of road embankments.

Six-lined racerunners are associated with dry, sunny habitats, where they are wide-foraging hunters.

Six-lined Racerunner
Aspidoscelis sexlineatus

FOOD AND FEEDING These wide-ranging, active foragers are always in search of insect prey. They have been recorded eating termites, caterpillars, butterflies, grasshoppers, and beetle larvae and adults. They will also eat spiders and may eat scorpions as some other whiptail lizards are known to do.

REPRODUCTION Six-lined racerunners mate in April and May in the Southeast. The underside of adult males becomes darker blue during the mating period when they are actively searching for females. Females lay two to eight eggs (average about three) in late May and June in a nest that is usually 4–8 inches deep in sand or soft soil. There are several reports of six-lined racerunners laying their eggs in sawdust piles. Most hatchlings emerge in July and August after an incubation period of up to 2 months; some may emerge as late as early September.

PREDATORS AND DEFENSE The known predators include black racers, coachwhips, common terns, and eastern glass lizards. The primary mode of defense is running rapidly for cover or losing part of the long tail. Introduced fire ants eat the eggs.

CONSERVATION ISSUES Six-lined racerunners and most of their habitats are not considered threatened in most parts of the Southeast, but loss of habitat and the application of pesticides that affect their prey could potentially pose problems for this species.

Did you know?

Some species of whiptail lizards (genus Aspidoscelis) in the U.S. Southwest produce offspring without mating, and no males have ever been found.

A longitudinal dark stripe down the back and dark stripes on each side of the body below the lateral fold are characteristic of the slender glass lizard.

How do you identify a slender glass lizard?

BODY PATTERN AND COLOR
brownish with dark longitudinal stripes on back and sides

DISTINCTIVE CHARACTERS
dark stripes down center of back and along body beneath lateral fold

SIZE

42" } 7"

● ADULT
● HATCHLING

Slender Glass Lizard   *Ophisaurus attenuatus*

FAMILY Anguidae

DESCRIPTION Slender glass lizards are brown to golden brown, bronze, or tan overall, with a central dark stripe extending from the head down the length of the back and becoming lighter on the tail. This striping is very dark on small to medium-sized specimens and usually becomes less distinct with increasing age and size. Old males may have a speckled "salt-and-pepper" coloration, and some adults have distinctive crossbars. Dark longitudinal stripes are present on each side below the lateral fold and under the tail. The belly is cream colored. Males get larger than females. More than 97 body scales are present along the lateral fold.

VARIATION AND TAXONOMIC ISSUES Slender glass lizards are subdivided into eastern and western subspecies: *O. a. longicaudus* and *O. a. attenuatus,* respectively. Both occur in southeastern states, although their geographic ranges are widely separated by the Mississippi River valley. The two subspecies vary somewhat in color throughout their ranges but without notable geographic trends. The eastern subspecies reaches a larger size and has a proportionately longer tail than the western. Some lizard biologists consider the eastern and western slender glass lizards to be distinct species.

WHAT DO THE HATCHLINGS LOOK LIKE? Hatchlings look like miniature adults, but they lack the brown crossbars present on some older lizards, and their stripes are darker (sometimes almost black) and bolder than those of adults. The hatchlings may have only a single dark stripe down the sides rather than several.

CONFUSING SPECIES The dark stripe down the back and the distinct dark longitudinal stripes on each side below the lateral fold will usually distinguish slender glass lizards from the other three species present in the Southeast. The mimic glass lizard has faint stripes below the lateral fold rather than dark ones. The slender glass lizard has more than 97 body scales along the lateral fold whereas the mimic and island glass lizards have fewer than 97.

DISTRIBUTION AND HABITAT The slender glass lizard is the most widespread of the four glass lizard species and occurs in all or part of every south-

eastern state, being absent from mountainous areas in North Carolina, Tennessee, and Virginia, much of the Mississippi River floodplain, and northeastern and central Louisiana. The eastern subspecies is found in every southeastern state; the western subspecies is found in the western portion of Louisiana. Dry, upland areas such as pine woods and sandhills in open habitats, including vacant lots, are preferred.

Slender glass lizards occupy open habitats in dry, upland areas, where they forage for insects among ground litter.

Slender Glass Lizard
Ophisaurus attenuatus

- eastern slender glass lizard
  O. a. longicaudus
- western slender glass lizard
  O. a. attenuatus

BEHAVIOR AND ACTIVITY Slender glass lizards are most active in the summer in the morning and late afternoon. They are more likely than the other species of glass lizards to thrash violently when captured, and many break their tail as a result. Also unlike the other glass lizards, slender glass lizards hide under low vegetation rather than burrowing in the sand during the warm parts of the year. They hide under tufts of grass or other patches of vegetation during much of the day but do go completely underground during hibernation.

FOOD AND FEEDING These active foragers feed on many invertebrates—including beetles, ants, spiders, caterpillars, and snails—and small lizards and snakes. Grasshoppers seem to make up a large part of the diet, probably because they are common in the open grassy areas where these lizards live.

REPRODUCTION Slender glass lizards in the Southeast mate in the spring, but little is known about their courtship and mating behavior. Females lay clutches of 4–19 eggs from late spring to midsummer in shallow depressions under vegetation or decaying logs. The female stays with the eggs and will gather them if they are scattered. The incubation period is about 60 days.

Slender glass lizards are most often found away from populous, urban areas.

PREDATORS AND DEFENSE Known predators include a variety of snakes—among them coachwhips, black racers, coral snakes, and kingsnakes—as well as raptors and predatory mammals.

CONSERVATION ISSUES Eastern slender glass lizards do not occur in high densities anywhere but are still found throughout their historical range in areas that are not highly urbanized. Their populations are unquestionably affected by habitat fragmentation and loss, and may have declined in some areas where pesticide use has reduced their potential prey.

The island glass lizard has a single dark stripe above the lateral fold.

Island Glass Lizard

Ophisaurus compressus

FAMILY Anguidae

DESCRIPTION Island glass lizards are brown to tan, and most have a dark longitudinal stripe or dark dashes down the middle of the back. A single dark longitudinal line is usually present along each side above the lateral fold. There are no stripes below the lateral fold. Many white marks are evident along the sides of the neck. The belly is typically pale yellow. In contrast to the other southeastern glass lizards, the tail vertebrae lack the fracture planes that allow the tail to break freely into sections. Individuals have fewer than 97 scales along the lateral fold.

VARIATION AND TAXONOMIC ISSUES No subspecies or geographic variation of any sort has been described.

WHAT DO THE HATCHLINGS LOOK LIKE? Hatchlings presumably resemble the adults.

CONFUSING SPECIES Island glass lizards, slender glass lizards, and mimic glass lizards all may have dark stripes down the middle of the back. The single dark stripe above the lateral fold in the island glass lizard is the key to telling the three apart. The slender glass lizard has dark stripes on the sides below the lateral groove, and the mimic glass lizard has several stripes both above and below the lateral fold.

How do you identify an island glass lizard?

BODY PATTERN AND COLOR brownish with a longitudinal dark stripe down each side

DISTINCTIVE CHARACTERS longitudinal dark stripes down the side above the lateral fold and sometimes a dark stripe down the back

SIZE

24" 5"

● ADULT
● HATCHLING

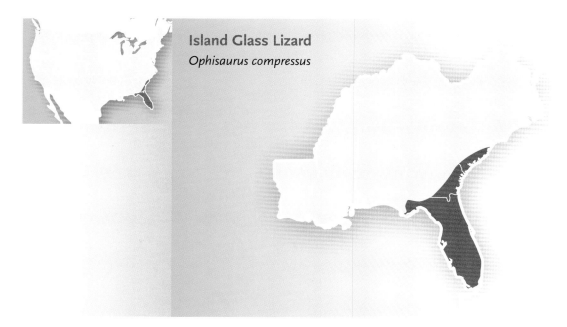

Island Glass Lizard
Ophisaurus compressus

Clumps of muhly grass on Little St. Simons Island, Georgia, are typical habitat of the island glass lizard.

DISTRIBUTION AND HABITAT The island glass lizard is found throughout peninsular Florida, in the Florida Panhandle west to Apalachicola Bay, and in a narrow band of coastal habitat in South Carolina and Georgia. The preferred habitat is primarily sandy, including areas of beach vegetation and tidal wrack above the surf line. Specimens are most likely to be found under vegetative and other debris.

BEHAVIOR AND ACTIVITY The island glass lizard is active during the day, but little else is known about the specific ecology and behavior of this secretive lizard.

FOOD AND FEEDING Presumably the diet includes insects, spiders, and other small invertebrates.

REPRODUCTION Mating occurs in the spring. Males have been observed biting females behind the head and arching their body to the side to mate, with the pair remaining together for several hours. The female lays 4–18 eggs about a month and a half later and stays with them until they hatch. Incubation takes about 6 weeks.

PREDATORS AND DEFENSE Snakes, larger glass lizards, birds, and carnivorous mammals are the most likely predators of island glass lizards.

CONSERVATION ISSUES Although some of the coastal areas where this species is most common are protected, the rapid coastal development occurring throughout this lizard's range is undoubtedly affecting populations.

The mimic glass lizard is known to eat insects and spiders, but relatively little else is known of the ecology of this secretive species.

How do you identify a mimic glass lizard?

Mimic Glass Lizard   *Ophisaurus mimicus*

FAMILY Anguidae

DESCRIPTION The mimic glass lizard was not described as a separate species until 1987. The distinction was based in part on body scale counts on museum specimens, most of which had been wrongly identified as other species. The mimic glass lizard has fewer than 97 scales along the lateral fold. The general body color is brown or tan, and a dark longitudinal stripe is usually evident down the middle of the back. The stripe is typically darker toward the tail. Most individuals have a total of three or four longitudinal stripes above and below the lateral fold. These may be distinct stripes or a series of dashes or spots, and those below the lateral fold are faint. Some adults exhibit the crossbar pattern seen in slender glass lizards.

VARIATION AND TAXONOMIC ISSUES No subspecies or geographic variation has been described.

WHAT DO THE HATCHLINGS LOOK LIKE? Hatchlings have not been described. They probably resemble the adults but with more distinct markings.

CONFUSING SPECIES Adult mimic glass lizards are much smaller than adult slender or eastern glass lizards. Mimic glass lizards and slender glass

BODY PATTERN AND COLOR brownish with dark brown longitudinal stripe down center of back; 3–4 longitudinal stripes below the lateral fold are less distinct than in the slender glass lizard

DISTINCTIVE CHARACTERS center dark stripe and several dark stripes on sides, narrow stripes on sides of tail

SIZE

24"   5"

● ADULT
● HATCHLING

lizards may also have stripes both above and below the lateral fold, but the stripes below the groove on mimic glass lizards are much less distinct than those found on slender glass lizards. The eastern glass lizard has no stripe down the middle of the back. The island glass lizard does have a stripe down the middle of the back, but it also has a single stripe above the lateral groove and no stripes below it.

DISTRIBUTION AND HABITAT The mimic glass lizard has been found only in a narrow, crescent-shaped strip of coastal land extending from North Carolina around to Mississippi, being absent from most of the Florida peninsula. Some herpetologists believe that the species was associated with the more mesic portions of the longleaf pine–wiregrass community that historically covered much of the Southeast but is now rare. Specimens have been found in pitcher plant bogs and pine flatwoods in some parts of the range. This is decidedly not a sandhills species.

BEHAVIOR AND ACTIVITY Little is known about the ecology and behavior of the mimic glass lizard.

FOOD AND FEEDING The diet is presumed to be mostly invertebrates, including insects and spiders.

REPRODUCTION Courtship, mating, and nesting behavior have not been described.

Mimic Glass Lizard
Ophisaurus mimicus

PREDATORS AND DEFENSE Black racers are the only documented predator of the mimic glass lizard; potential predators include other snakes, larger glass lizards, birds, and carnivorous mammals.

CONSERVATION ISSUES Too little is known about the biology and ecology of the mimic glass lizard to determine its current status. However, the same issues affecting other species of glass lizards (and reptiles in general), including but not limited to habitat loss and fragmentation, could undoubtedly affect mimic glass lizard populations.

Mimic glass lizards lay eggs like other native southeastern species, but courtship, mating, and nesting behavior have not been described.

Adult male eastern glass lizards are often greenish in color.

How do you identify an eastern glass lizard?

BODY PATTERN AND COLOR
brown or greenish above with greenish white speckles on the sides; longitudinal dark stripes usually present on sides, but only above the lateral fold

DISTINCTIVE CHARACTERS
old adults are often greenish with yellow belly

SIZE

30" 7"

● ADULT
● HATCHLING

Eastern Glass Lizard *Ophisaurus ventralis*

FAMILY Anguidae

DESCRIPTION Eastern glass lizards have no dark stripe down the center of the back and may lack dark stripes altogether. Adult males often are greenish overall with white dots giving them a vivid speckled appearance; younger individuals and females may have dark longitudinal side stripes, but not below the lateral fold. White dash marks are evident on the sides of the neck of juveniles and some adults. The belly is yellow to cream. Many specimens have a distinctive regenerated tail tip that is plain brown. More than 97 body scales are present along the lateral fold.

VARIATION AND TAXONOMIC ISSUES No subspecies have been described and no geographic variation in color or body morphology has been noted.

WHAT DO THE HATCHLINGS LOOK LIKE? Hatchlings are beige to khaki colored, some with a greenish brown tint, and have a longitudinal dark stripe down each side.

CONFUSING SPECIES The lack of a dark stripe down the center of the back distinguishes eastern glass lizards from slender glass lizards and mimic glass lizards. The island glass lizard has one distinct stripe above the lateral fold and is immaculate below it, while the eastern glass lizard may have several less distinct stripes above the fold. In addition, the eastern gets much larger than the largest island glass lizard. The presence of more than 97 body scales along the lateral fold distinguishes eastern glass lizards from mimic and island glass lizards, both of which have fewer.

DISTRIBUTION AND HABITAT Eastern glass lizards occur throughout the Coastal Plain and in some of the Piedmont from southern Virginia to eastern Louisiana. They occur in pine flatwoods but are also common in moister habitats, including the edges of freshwater wetlands, beaches, and salt marshes. They are especially common on some barrier islands in open grassy habitat and along road shoulders.

BEHAVIOR AND ACTIVITY Eastern glass lizards forage aboveground during the daytime, especially in the morning, and are active burrowers in sand and loose soil. They have been observed at night during periods of heavy rain, possibly after being flooded out of a refuge. Eastern glass lizards are usually calmer when handled than the more excitable slender glass lizard.

FOOD AND FEEDING The diet includes a variety of insects, spiders, snails, and even small crabs. These lizards are also known to eat small snakes, lizard eggs, eggs of ground-nesting birds, and the young of small mammals. They are cannibalistic, too, eating smaller individuals of their own species as well as other small lizards.

Did you know?

Most species of North American lizards have four legs, but one has only two, and several have no legs at all.

Eastern Glass Lizard
Ophisaurus ventralis

The eastern glass lizard is a daytime visual hunter of insects and other small prey.

REPRODUCTION Mating presumably occurs in the early spring, but little is known about the mating behavior. Clutches of 4–15 eggs have been documented. Eggs are usually laid in late spring to early summer in protected depressions such as beneath clumps of vegetation, logs, or man-made debris. The female exhibits maternal behavior by staying with her eggs until they hatch after about 60 days, usually in August or September. She may not actively defend the eggs but will regather any that become separated from the others.

PREDATORS AND DEFENSE Snakes, hawks and other birds, and carnivorous mammals are predators of eastern glass lizards.

CONSERVATION ISSUES No specific threats to this species have been identified aside from the widespread development that can destroy their natural habitat.

Little brown skinks do not climb trees and are nearly always found on the ground or beneath leaves, logs, and boards.

Little Brown Skink

Scincella lateralis

FAMILY Scincidae

DESCRIPTION This very small, slender skink has a round body and an unusually long tail. Individuals vary in color from light tan to dark coppery brown with a wide stripe running down the back that may not be obvious on darker individuals. The belly varies from white to yellow. The tail is usually lighter than the body and varies from brown to gray. The legs are very small, and the head is narrower than that of other skinks.

VARIATION AND TAXONOMIC ISSUES Another common name is ground skink. No subspecies are documented, and notable geographic variation has not been reported.

WHAT DO THE HATCHLINGS LOOK LIKE? The young look like adults and average about 1.75 inches at hatching.

CONFUSING SPECIES Little brown skinks may be confused with mole skinks, which are similar in size but have a lighter stripe above each eye and usually have a blue or red tail. Sand skinks have much smaller legs and only one or two toes per foot.

How do you identify a little brown skink?

BODY PATTERN AND COLOR small, coppery colored, with a wide darker stripe running down the middle of the back; belly white to yellow

DISTINCTIVE CHARACTERS window in lower eyelid permits vision when eyes are closed

SIZE

5"   1.75"

● ADULT
● HATCHLING

The little brown skink lacks the pronounced yellow stripes exhibited by many of the other skinks with which its range overlaps.

DISTRIBUTION AND HABITAT Little brown skinks occur throughout all or most of each southeastern state but are not common at higher elevations. They are relatively ubiquitous in wooded habitats, including pine and hardwoods, moist or dry. In urban and suburban areas they are often common in vacant lots, alongside buildings, and under board piles.

BEHAVIOR AND ACTIVITY These little lizards spend the majority of their lives in the leaf litter, basking or foraging aboveground and dashing beneath cover to avoid predators, the body thrashing from side to side as it disappears into the leaf litter. Locomotion is serpentine, as befits an animal with such a long tail and short legs. They can be active in any month, including winter, if temperatures are warm. Researchers at the Savannah River Ecology Laboratory in South Carolina have collected and released large numbers (as many as 50 in an hour) that were captured in drift fences and under coverboards in pine forests.

FOOD AND FEEDING Little brown skinks forage through leaves and other debris in search of small invertebrates, including small insects such as termites and isopods as well as appropriately sized spiders and millipedes. They have been observed to even bite their own tail, presumably mistaking it for prey.

REPRODUCTION Little brown skinks mate in the spring. Between April and July females lay 1–7 (average 2–3) eggs out of sight in decaying wood and vegetation. They may lay up to four clutches per year, about 3–4 weeks apart. Unlike many other southeastern skinks, little brown skink females do not remain with their eggs. More than one female will lay eggs in the same nest, however, the record being 66 eggs in one communal nesting site. Most of the young emerge from the nest in July and August.

PREDATORS AND DEFENSE Many snakes, including common and scarlet kingsnakes, copperheads, and ringneck snakes, are known predators, as are wolf spiders, bluebirds, and barred owls. Many other birds and small mammals will eat a small lizard if given the opportunity. Little brown skinks frequently fall victim to domesticated cats and dogs. The typical escape pattern is to retreat beneath leaves, pine straw, logs, or other objects, but these lizards are known to run into standing water and swim to the other side or even to stay in the water with only the head visible. They do not try to escape into trees the way some of the larger skinks do.

CONSERVATION ISSUES This species has a large geographic range and uses a wide variety of habitats, and no generalized threats are recognized.

Little Brown Skink
Scincella lateralis

The coal skink has a more fragmented geographic distribution than any other native species of southeastern lizard.

How do you identify a coal skink?

BODY PATTERN AND COLOR
brown with a wide, darker band running the length of the body on each side

DISTINCTIVE CHARACTERS
dark band on body bordered by two thin light stripes down each side.

SIZE

7"   2"

● ADULT
● HATCHLING

Coal Skink

Plestiodon anthracinus

FAMILY Scincidae

DESCRIPTION Coal skinks are brown, medium-sized (maximum total length about 7 inches) lizards with smooth, shiny, overlapping scales. A distinctively dark stripe that is up to four body scales wide runs the entire length of the body on each side. Each dark stripe is framed by a pair of thin pale stripes that also run along the body onto the tail. The belly is dark. Adult males and females are similar in appearance except during the breeding season, when the head of males turns reddish orange along the sides.

VARIATION AND TAXONOMIC ISSUES Two subspecies are recognized—the northern coal skink (*P. a. anthracinus*) and the southern coal skink (*P. a. pluvialis*)—and both are found in the Southeast. Adults of the two subspecies do not differ appreciably in general appearance. In some areas coal skinks are called black skinks. *Plestiodon* replaces the former genus name, *Eumeces*.

Juveniles of the southern subspecies of coal skink are black or dark blue with no stripes.

Both subspecies (northern, left, and southern, right) of the coal skink are found in the Southeast.

WHAT DO THE HATCHLINGS LOOK LIKE? Northern coal skink hatchlings look like the adults, with visible stripes. Southern coal skink hatchlings are dark blue or black without obvious stripes. The tail color of young skinks varies from iridescent violet to blue or black, depending on the geographic region and the age of the individual.

CONFUSING SPECIES Several species of dark skinks in the Southeast have longitudinal lines and are similar in appearance. Adult coal skinks are the only skinks with a wide, dark line down each side of the body that is more than two (up to four) body scales wide as well as only four light-colored lines that run the length of the body. The presence of two small scales on the underside of the lower jaw distinguishes the coal skink (which has only one scale) from the other skinks within its geographic range.

DISTRIBUTION AND HABITAT The coal skink has been reported from every southeastern state, but the distribution pattern is irregular. Great distances often separate isolated concentrations of populations. The habitat is typically a heavily shaded, moist hillside in a deciduous or mixed hardwood-pine forest, often in association with rocky areas near water.

BEHAVIOR AND ACTIVITY Less is known about the ecology and behavior of coal skinks than about most of the other lizards native to the Southeast. The scattered distribution patterns and small population sizes where they

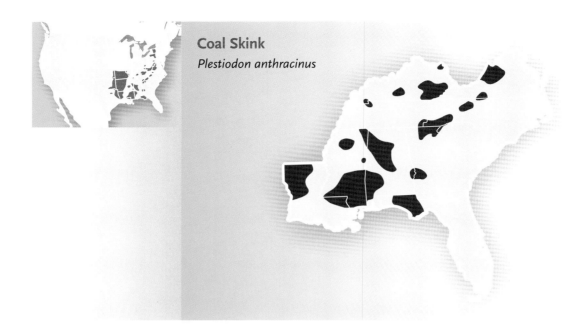

Coal Skink
Plestiodon anthracinus

are present make study difficult. Coal skinks differ from the common five-lined skinks in being more terrestrial and less likely to climb trees. They are often found underground. They are most active during late winter and early spring but may be seen at any time of the year.

FOOD AND FEEDING Coal skinks eat small invertebrates, including insects (beetles, termites, and grasshoppers), spiders, and earthworms, which they locate in leaves and other ground litter.

REPRODUCTION Courtship and mating occur in February or March. Females lay 4–11 eggs (average about 6 or 7) between April and June. The females guard the eggs, which take 4–5 weeks to hatch.

PREDATORS AND DEFENSE Probable predators of coal skinks in areas of overlap are kingsnakes, racers, copperheads, and birds of prey. Coal skinks living around wetlands may run into water when pursued and swim to the bottom. If rocks are present in the water, the skink will usually hide beneath them. Coal skinks living in deciduous forests disappear quickly underground when threatened.

CONSERVATION ISSUES The coal skink is so uncommon throughout its southeastern geographic range that no clear conservation strategies have been developed. Presumably, protecting the moist forests where populations of the species occur is an important consideration.

Mole skinks are secretive lizards that hide under sand or ground litter. The Florida Keys subspecies shown here has a reddish tail.

Mole Skink

Plestiodon egregius

FAMILY Scincidae

DESCRIPTION These secretive little lizards are rarely more than 5 inches long, with a very slender body and minuscule limbs. Their shiny scales give them a moist appearance. They show remarkable variation in color throughout their range. The body coloration varies from light gray to dark brown. Some individuals have vivid stripes down the body while others are almost immaculate. The tail color is highly variable both among individuals and among populations, and ranges from yellow to red to even violet or blue in the blue-tailed subspecies. Females grow slightly larger than males. During the breeding season, males develop a light prenuptial blush of red or orange on the sides of the head and lips.

VARIATION AND TAXONOMIC ISSUES Five subspecies are recognized: Florida Keys mole skink (*P. e. egregius*), northern mole skink (*P. e. similis*), blue-tailed mole skink (*P. e. lividus*), Cedar Key mole skink (*P. e. insularis*), and peninsula mole skink (*P. e. onocrepis*). The color pattern varies widely across the geographic range. *Plestiodon* replaces the former genus name, *Eumeces*.

WHAT DO THE HATCHLINGS LOOK LIKE? The hatchlings are about 2 inches long and usually look very similar to adults. Hatchlings of the Cedar Key subspecies are almost immaculate black.

How do you identify a mole skink?

BODY PATTERN AND COLOR
small, slender, with very small limbs; body gray to tan; tail orange, red, lavender, or blue

DISTINCTIVE CHARACTERS
two lighter stripes down the sides of the body, their length varying between individuals; light stripe evident above each eye

SIZE

5"   2"

● ADULT
● HATCHLING

The northern subspecies of the mole skink usually has light stripes on the sides and a tail that is reddish. This specimen is from Long County, Georgia.

The subspecies known as the peninsula mole skink is highly variable in tail color, ranging from reddish brown to orange, pinkish, and lavender.

The Cedar Key mole skink subspecies is confined to Levy County, Florida, on Cedar Key and Seahorse Key.

The blue-tailed mole skink occupies a restricted geographic range along the sand ridge of central Florida.

Mole Skink
Plestiodon egregius

northern mole skink
P. e. similis

peninsula mole skink
P. e. onocrepis

Cedar Key mole skink
P. e. insularis

blue-tailed mole skink
P. e. lividus

Florida Keys mole skink
P. e. egregius

CONFUSING SPECIES Mole skinks might be confused with little brown skinks where their ranges overlap, but the latter have a brown tail. Mole skinks are smaller than adults of most other skink species and more slender. The five-lined skinks have distinct lines on the back and a bright blue tail as juveniles. Florida sand skinks have proportionately smaller front legs, no ear openings, and a lighter body coloration.

DISTRIBUTION AND HABITAT Mole skinks range through portions of Alabama, most of Florida, and the lower half of Georgia. They occur along the Georgia–South Carolina border but have not been recorded across the Savannah River in South Carolina. They live in dry woodland habitats—including longleaf pine–turkey oak, sandy pine uplands, and scrub—and are excellent burrowers, seeming to dive directly into the ground when threatened. They can also be found near the ocean and in salt marsh habitat beneath tidal wrack and other debris. The Florida Keys subspecies is found on the Dry Tortugas.

BEHAVIOR AND ACTIVITY These secretive little lizards are particularly adept at disappearing into the sandy substrate when disturbed, and spend most of their time either underground or under some form of cover, most commonly rocks, logs, palm fronds, or other debris. Specimens may also be collected by raking through the mounds created by pocket gophers, which share much of the mole skink's range. The skinks burrow into these sandy mounds in late winter and early spring, presumably because they can stay warmer.

Mole skinks have tiny legs and slender bodies.

FOOD AND FEEDING Mole skinks feed on a variety of small arthropod prey, including insects, especially crickets, as well as spiders, and even crustaceans in certain parts of their range.

REPRODUCTION Female mole skinks lay two to nine eggs in April through June in a nest they excavate. The nest may be a few inches below the soil surface to as many several feet underground. The female attends the eggs during the 4.5 to 7-week incubation period, presumably to guard them from potential predators.

PREDATORS AND DEFENSE Mole skinks make ideal fare for scarlet kingsnakes and other small snakes that share their habitat; birds and small mammals are also among their predators.

CONSERVATION ISSUES Conservation efforts need to focus on the habitat instead of the individual species. Mole skinks inhabit upland areas that are particularly vulnerable to the pressures of residential and commercial development.

A breeding male common five-lined skink has developed red coloration on the head while still retaining a bluish tint on the tail.

How do you identify a common five-lined skink?

Common Five-lined Skink   *Plestiodon fasciatus*

FAMILY Scincidae

DESCRIPTION Like the southeastern five-lined and broad-headed skinks, these medium-sized lizards are often called scorpions in the mistaken belief that they are venomous. Large adult males may be about 8 inches long with a snout–vent length of about 3 inches. Males are slightly larger than females.

VARIATION AND TAXONOMIC ISSUES Common five-lined skinks do not vary noticeably across their geographic range, and no subspecies are recognized.

WHAT DO THE HATCHLINGS LOOK LIKE? The hatchlings are stunning animals with a dark blue-black body and bold yellowish stripes. They have a blue tail that in young juveniles appears bright cobalt. *Plestiodon* replaces the former genus name, *Eumeces*.

This juvenile common five-lined skink has the species' signature yellow stripes and bright blue tail.

BODY PATTERN AND COLOR
females and juveniles dark blue-black with conspicuous stripes running down the body; older males fade to immaculate brown, often with a reddish head

DISTINCTIVE CHARACTERS
central row of scales beneath tail wider than adjacent ones; four lip scales

SIZE

8"   2"

● ADULT
● HATCHLING

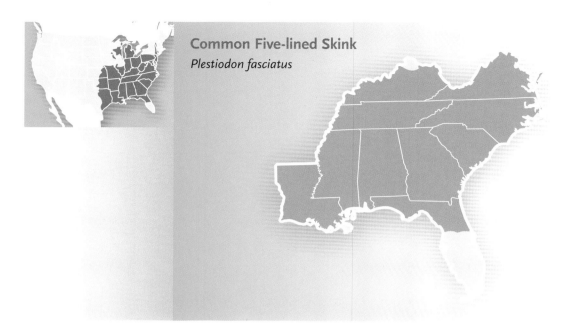

Common Five-lined Skink
Plestiodon fasciatus

As common five-lined skinks mature, the tail color fades from blue to dull brown.

CONFUSING SPECIES Common five-lined skinks are easily confused with other skink species, especially the southeastern five-lined skink and the broad-headed skink (see the broad-headed skink account, page 105). The coal skink account (page 92) describes how to differentiate between that species and the common five-lined skink.

The row of broader scales down the center of the underside of the tail of the common five-lined skink and the broad-headed skink distinguish the two species from the southeastern five-lined skink.

DISTRIBUTION AND HABITAT This extremely wide-ranging species occurs in every state in the southeastern United States but is not found in the lower half of Florida. Common five-lined skinks are common in mesic forests along stream banks and other sources of water. They frequent a variety of damp woodland habitats, especially those strewn with rotting logs and other debris. They are both terrestrial and arboreal.

BEHAVIOR AND ACTIVITY These conspicuous lizards are often seen basking on the ground at the base of trees or stumps, and on fallen logs.

FOOD AND FEEDING The diet consists almost exclusively of arthropods, including a variety of insects and spiders. Common five-lined skinks often forage through leaf litter searching for appropriately sized invertebrate prey.

REPRODUCTION Females lay 4 to 15 eggs between April and June. The female broods her eggs (presumably guarding them from predation) during the 4 to 8-week incubation period and may assist in the hatching process. While they are incubating she rotates and turns the eggs and may even defecate on them to provide additional moisture if they start to dry out.

A female common five-lined skink broods her eggs.

PREDATORS AND DEFENSE Known predators include a variety of snakes, birds, dogs, cats, and other predatory mammals. A hatchling was observed wrapped up in silk in a spiderweb.

CONSERVATION ISSUES The real threat to skinks is habitat destruction. Many of the forested habitats in which they live have been cleared and developed into residential neighborhoods or commercial areas. Other potential problems for skink populations include cats and other domestic pets that may prey on them and the use of herbicides and pesticides, which may affect them directly or reduce their available food.

Most southeastern five-lined skinks have yellowish or orange stripes and a bright blue tail.

How do you identify a southeastern five-lined skink?

BODY PATTERN AND COLOR
females and juveniles dark blue or black with conspicuous stripes running the length of the body; older males faded, immaculate brown, often with a reddish head; tail of young juveniles bright cobalt blue

DISTINCTIVE CHARACTERS
all scales beneath tail approximately the same size

SIZE

6"   2"

● ADULT
● HATCHLING

Southeastern Five-lined Skink

Plestiodon inexpectatus

FAMILY Scincidae

DESCRIPTION These medium-sized lizards look very much like the other two large skinks found in the Southeast. They are shiny lizards with smooth scales and a small head. Juveniles and females have five bold stripes running down the body, two faint stripes on either side of the belly, and an electric blue tail. Adult males usually fade to tan or brown with age but have a bright orange head during the breeding season. Adult females often retain the bold stripes.

VARIATION AND TAXONOMIC ISSUES No subspecies are recognized. *Plestiodon* replaces the former genus name, *Eumeces*.

WHAT DO THE HATCHLINGS LOOK LIKE? Newly hatched southeastern five-lined skinks are boldly patterned with white or yellow stripes running the length of the dark body and a bright blue tail.

CONFUSING SPECIES Southeastern five-lined skinks can easily be confused with common five-lined skinks and broad-headed skinks; see the broad-headed skink (page 105) and coal skink (page 92) accounts for identification clues.

DISTRIBUTION AND HABITAT This species is found in at least part of every southeastern state from eastern Virginia to easternmost Louisiana and into most of Tennessee and southern Kentucky. Southeastern five-lined skinks occupy a variety of dry habitats including sandhills, hickory forests, and sea islands along the Gulf and Atlantic coasts. They are particularly at home in areas with rotten logs and other debris in which to hide and forage.

BEHAVIOR AND ACTIVITY These skinks bask or prowl through the leaf litter in search of suitable prey during the day, quickly dashing under cover at the slightest threat. They spend nights and periods of inactivity under logs and other debris.

Females of many skink species brood their eggs and protect them from small predators.

FOOD AND FEEDING Southeastern five-lined skinks feed on a variety of arthropods, including insects and spiders.

REPRODUCTION Increased testosterone levels during the spring mating season cause the males to develop a reddish head. The female lays 4–11 eggs in a rotten log, stump, or other cavity during June or July. The hatchlings emerge from the nest in about 4–6 weeks, with the incubation period depending on the nest temperature.

PREDATORS AND DEFENSE Like other skinks, juveniles have a bright blue tail that many scientists believe warns potential

Southeastern Five-lined Skink
Plestiodon inexpectatus

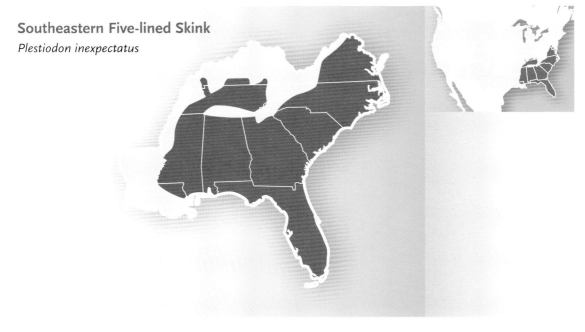

The tail of the southeastern five-lined skink is sometimes purplish. This individual is from the Savannah River Site in South Carolina.

predators that the skink is unpalatable. Known predators include a variety of snakes, birds, dogs, cats, and other predatory mammals, however, so either the skinks are not toxic or predators must sample one to learn that the taste and side effects are unpleasant.

CONSERVATION ISSUES The primary threat to southeastern five-lined skinks is habitat destruction for residential neighborhoods or commercial areas. Other potential problems for skink populations in suburban areas include predation by cats and other domestic pets. Herbicides and pesticides may affect skinks directly or indirectly by reducing the food available to them.

Broad-headed skinks, like this male basking in a neighborhood in Aiken, South Carolina, are among the most commonly seen large lizards native to the Southeast.

How do you identify a broad-headed skink?

Broad-headed Skink

Plestiodon laticeps

FAMILY Scincidae

DESCRIPTION Broad-headed skinks are the largest native lizards (with legs) in the Southeast. The maximum total body length recorded for an individual is slightly less than 13 inches. The body is wide, thick, and elongated, and the legs are relatively short. The thick, rounded tongue is notched on the end. Adult males are bigger than females and have an immaculate olive or brown body, wide jaws (or jowls), and a robust head that turns bright red or orange during the breeding season (April–June). Younger males often have faint stripes down the back. The slightly smaller females have a smaller head relative to their body size and varying degrees of striping down the back.

When not in breeding colors, the male (right) can be distinguished from the female because of its proportionately larger head. This older female (left) has lost much of the striping visible on smaller individuals.

BODY PATTERN AND COLOR smooth, overlapping scales very shiny, giving a wet, glasslike appearance; young females and juveniles with blue tail and light stripes on dark body; adult males brownish with reddish head

DISTINCTIVE CHARACTERS central row of scales beneath tail wider than adjacent ones; five lip scales

SIZE

9"   3"

● ADULT
● HATCHLING

VARIATION AND TAXONOMIC ISSUES Broad-headed skinks do not vary noticeably across their southeastern range. Some herpetologists have noted variation in scale characters and other features between eastern and western populations, but no subspecies are recognized in the Southeast. Local common names include scorpion and copperhead. *Plestiodon* was proposed to replace the former genus name for all North American lizards in the genus *Eumeces* on the basis of genetic studies undertaken in the late 1900s.

WHAT DO THE HATCHLINGS LOOK LIKE? Baby "broadheads" are stunning lizards with a shiny black body and five bold yellow or orange stripes running their entire length. The tail is electric blue. A person unfamiliar with native lizards might easily mistake an immobile specimen for a brightly colored plastic toy. The colors gradually fade with maturity, although adult females often retain some of the juvenile pattern.

CONFUSING SPECIES Their large size is often sufficient to distinguish adult broad-headed skinks of both sexes from other skinks of the Southeast. Young adults and juveniles might easily be confused with other species, however, especially the common five-lined skink and the southeastern five-lined skink. Identification often requires capture and careful examination. The scales beneath the tail of the southeastern five-lined skink are all the same size, while both broad-headed skinks and common five-lined skinks have a central row of wider scales down the underside of the tail. Distinguishing a broad-headed skink from a five-lined skink is often more

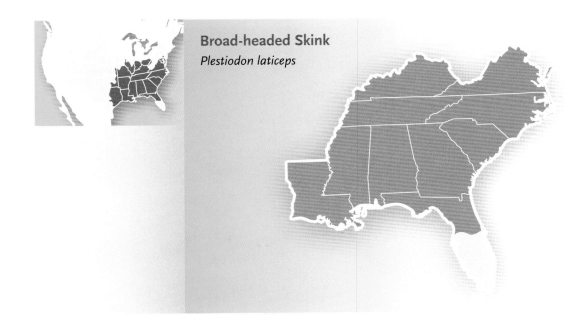

Broad-headed Skink
Plestiodon laticeps

difficult. Although not always reliable, two scale characters typically distinguish the two. The five-lined skink usually has two small scales touching the front of the ear, whereas the broad-headed skink usually has none. Also, the broad-headed skink usually has five labial (lip) scales preceding the scale beneath the eye while the common five-lined skink has only four. The coal skink account (page 92) describes how to distinguish the broad-headed skink from that species.

DISTRIBUTION AND HABITAT Habitats include a variety of woodland areas ranging from cypress swamps and moist hardwood forests to upland hickory forests and pine stands. One important component is the presence of large hollow trees, rotting hardwood logs, or other debris that can be used for cover. The geographic range includes all or most of the southeastern states, the exceptions being high-elevation areas of eastern Tennessee and western Virginia, and the southern half of Florida.

BEHAVIOR AND ACTIVITY Broad-headed skinks have been the subject of extensive field and laboratory research. Much of what has been learned about them is probably applicable to other skink species as well. "Broadheads" become active in late March or April, the exact time depending on the temperature and region. Although they sometimes bask and forage in leaf litter on the ground, they prefer an arboreal existence. When alarmed, a broad-headed skink will often scurry to the top of a tree. Abandoned cavities, knotholes, or fissures in the bark provide refuge from predators and a place to stay moist during dry conditions, as do cracks and crevices in old barns and outbuildings. These alert and wary lizards are difficult to capture; the smooth, muscular body is rather slippery when grasped. Larger individuals have strong jaws and may bite hard and hold on if handled. They can deliver a painful pinch, but their bites seldom break the skin. This is a long-lived species; individuals are known to have survived for at least 8 years.

In many species of lizards, adult males and females differ in color pattern and body size. In broad-headed skinks the female (top) often retains more of the juvenile colors, whereas the male (bottom) gets larger and develops a brown body and reddish head.

FOOD AND FEEDING Broad-headed skinks forage actively, tongue-flicking and digging through leaf litter to locate their prey, which they grasp, crush in their powerful jaws, and swallow whole. They use their tongue and vomeronasal system to discriminate among food choices, selecting foods they like to eat and avoiding those they do not like. The diet includes a wide variety of invertebrates, and large adults are capable of consuming small vertebrates as well, including young mice, nestling birds, and other lizards. They will also eat fruits such as blackberries, grapes, and mulberries. Broad-headed skinks have been observed raiding and even shaking paper wasp nests in order to eat the larvae. They will not eat large female velvet ants—wingless wasps locally known as "cow killers" whose irritant hairs, strong jaws, powerful sting, and remarkably hard exoskeleton make them very unattractive as prey.

REPRODUCTION Mating takes place in the spring (April and May), and each female usually lays a single clutch of eggs. Males emerge from brumation with a tan head that becomes bright orange or red when they enter breeding condition. The release of testosterone responsible for the color change also increases the acuity of the vomeronasal system, allowing males to identify females and determine whether they are receptive to mating. During the breeding season, males size each other up, determining dominance by grappling with their jaws. The smaller male usually retreats, but if the two are similar in size, savage fighting over territory and females will ensue. It is not uncommon to witness two males locked in combat, jaws clamped together and bodies entwined. Adult males often sport battle wounds or scars from prior skirmishes with rivals. A male may sometimes lose its tail during a fight, a biologically costly event because the tail stores energy in the form of fat. By midsummer, after the mating season is over, the males' bright head color has faded back to tan and they are no longer aggressive toward each other.

Females excrete pheromones from glands around the cloaca to let males know they are receptive to mating. Males locate receptive females by tongue-flicking the substrate. When a male finds a female ready to breed, he grasps her by the neck and copulates with her. He may remain nearby for several days to keep her from mating with other males. Soon after mating, in the late spring or early summer, females lay clutches of 6–16 eggs; larger females lay more eggs than smaller ones. Common nest sites are hollow logs, cavities in tree trunks, and under hardwood logs or bark on the ground; other sites include refuse mounds such as sawdust piles and mulch piles. The female often pushes debris around the eggs to protect them and keep them moist. She will remain with the eggs, presumably to protect them from scarlet snakes, ants, and other predators that would consider the tiny eggs a nutritious and tasty meal. The female may move the eggs around

Female broad-headed skinks lay clutches of 6 to 16 eggs, with larger females, such as the one shown here, laying more eggs than smaller ones.

within the nest to keep them moist, and in an unusual example of maternal "nursery maintenance" will even eat damaged eggs that could spoil the nest with bacteria or fungus. The boldly patterned hatchlings emerge from the eggs 25–50 days later.

PREDATORS AND DEFENSE Skinks fall prey to a variety of native mammalian predators and snakes, especially coachwhip snakes, racers, and kingsnakes. Juvenile skinks are eaten by scarlet kingsnakes and other small snakes as well as by garden spiders. Domestic cats are common predators. Injured adults can become victims of parasitic blowflies, which lay their eggs in the wound.

Tongue-flicking allows skinks to detect and avoid predators. The bright blue tail of juvenile skinks is assumed to protect them from predators in some way, but the mechanism has not been determined with certainty. The color may be a warning sign to predators that the lizard is poisonous, as some skinks are reported to be; or perhaps predators are more likely to focus on and strike the colorful tail than the body.

Juvenile skinks not only regenerate lost tails, they are able to catch up in terms of total body growth with skinks that have not lost their tails. Thus, the only real cost to juveniles for losing a tail is an increased vulnerability to predation while they regrow it. Tail loss for an adult can have more serious consequences because the tail stores body fat that is used for energy. For females, losing a tail can result in a smaller clutch size; for males, it may affect social status.

CONSERVATION ISSUES Unfortunately, the inoffensive broad-headed skink has an undeserved reputation for being venomous—perhaps because its broad head resembles the head of a venomous snake—and is widely feared and disliked. The common names scorpion and copperhead are misleading and inaccurate. Uninformed people, fearing their children or pets are in danger, kill these harmless reptiles instead of using them to teach their children to appreciate native wildlife. The real threat to skinks, however, is loss of habitat, including tree removal and activities associated with urbanization. The forested habitats in which they live are being cleared and developed into residential neighborhoods or commercial areas at an alarming rate. Not only is valuable habitat lost, but such developments introduce cats and other domestic pets that prey on skinks and can extirpate populations. In addition, herbicides and pesticides both poison skinks that consume them and reduce the skinks' available food. In some instances, chemicals ingested by skinks have been reported to disrupt hormonal activity necessary for successful reproduction.

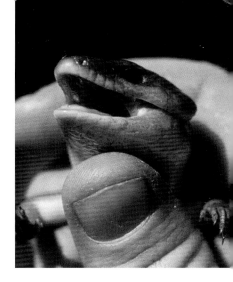

The broad-headed skink (shown here) usually has five labial (lip) scales preceding the scale beneath the eye, a trait that distinguishes it from the common five-lined skink, which has only four.

The Florida sand skink, which has a tiny pair of front legs, spends most of its time underground in sandy areas.

How do you identify a Florida sand skink?

BODY PATTERN AND COLOR
basically gray, beige, or tan

DISTINCTIVE CHARACTERS
limbs tiny, front legs with only one toe, hind legs with only two toes; lack of ear openings; wedge-shaped snout

SIZE

5"   2.5"

● ADULT
● HATCHLING

Florida Sand Skink   *Plestiodon reynoldsi*

FAMILY Scincidae

DESCRIPTION These small, gray or light tan lizards have reduced limbs with only one toe on the front limb and two on the rear. The lower eyelid has a transparent window that allows light reception when the eyes are closed.

VARIATION AND TAXONOMIC ISSUES This unusual skink was once considered the only species in its own genus, *Neoseps*, with no subspecies or notable geographic variation. In 2005, herpetologists used molecular genetics techniques to determine that the Florida sand skink is closely related to the mole skink and subsequently placed both species in the genus *Plestiodon*.

WHAT DO THE HATCHLINGS LOOK LIKE? Hatchlings look like miniature adults with a dark band running down each side.

CONFUSING SPECIES Little brown skinks and mole skinks have larger front legs, ear openings, and darker body coloration.

DISTRIBUTION AND HABITAT Florida sand skinks are known from six counties along the major sand ridges in the middle of the Florida peninsula. The typical habitat is fire-maintained dry uplands with deep sands and

Florida Sand Skink
Plestiodon reynoldsi

sparse vegetation, including rosemary scrub and natural sand pine–scrub oak forests. Specimens are found less commonly in longleaf pine–turkey oak sandhills habitat and occasionally in citrus groves planted in former scrub oak forests and in scrubby flatwoods adjacent to scrub habitat. They sometimes burrow in the sand mounds created by pocket gophers.

BEHAVIOR AND ACTIVITY Florida sand skinks spend most of their time several inches to 2 feet or so beneath the sand, through which they move in a swimming motion. When they are moving near the surface, they sometimes leave an undulating trail in the sand. They are most active in seasons of moderate temperatures from spring to early summer and from late summer to early fall but may be active any time of the year, including winter if temperatures are sufficiently warm. Occasionally, individuals come above the surface but stay hidden beneath dead leaves, palmetto fronds, or other ground litter.

FOOD AND FEEDING Florida sand skinks capture termites, beetle grubs, and other insects underground or possibly on the surface beneath ground debris.

REPRODUCTION Florida sand skinks become reproductively active when they are 1.5–2 years old. Mating begins in late winter or early spring (mid-February to early May) depending on environmental temperatures. Females

When a Florida sand skink closes its eyes, it can detect light through a transparent area in the lower lid.

lay two or three eggs in May or June that hatch in July or August. Florida sand skinks are known to live for at least 8–10 years.

PREDATORS AND DEFENSE Predators probably include many of the snakes, other lizards, and birds, including Florida scrub-jays, that share their scrub habitat.

CONSERVATION ISSUES The Florida sand skink is listed by the federal government as a threatened species and is protected by the Endangered Species Act. The greatest threat to its existence is the destruction of the scrub habitat that it requires, most of which has already been developed for urban, suburban, or agricultural purposes. The intentional suppression of natural fires changes the presumably preferred open habitat to a canopy-covered forest that is too shady.

The narrow, dark longitudinal line down the sides distinguishes a prairie skink from other southeastern lizards.

Prairie Skink

Plestiodon septentrionalis

FAMILY Scincidae

DESCRIPTION These medium-sized lizards (up to 7 inches in total length) have a light to dark brown body and a black line on the upper sides that runs from the ear opening onto the tail. The dark line has a thin light-colored line above and below it. The scales are smooth and shiny.

VARIATION AND TAXONOMIC ISSUES Three subspecies have been described; the one found in the Southeast is the southern prairie skink, *P. s. obtusirostris*. *Plestiodon* replaces the former genus name, *Eumeces*.

WHAT DO THE HATCHLINGS LOOK LIKE? Hatchlings resemble the adults but have a blue tail.

CONFUSING SPECIES The dark line down the sides is narrower in the prairie skink than in the coal skink.

DISTRIBUTION AND HABITAT The southern prairie skink barely qualifies as a member of the southeastern lizard fauna by occurring sparsely in the two northwesternmost parishes in Louisiana. Habitats include open pastures, agricultural fields, rocky terrain, prairies, and dry woodlands.

How do you identify a prairie skink?

BODY PATTERN AND COLOR
back olive to brownish with stripes on the sides; gray belly

DISTINCTIVE CHARACTERS
dark stripe bordered by lighter stripes on sides running from head onto tail

SIZE

6"   1"

● ADULT
● HATCHLING

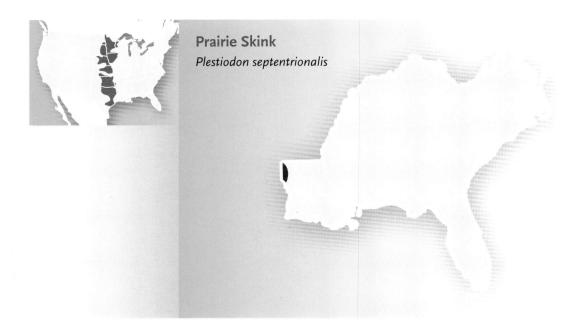

Prairie Skink
Plestiodon septentrionalis

BEHAVIOR AND ACTIVITY Little is known about the ecology of southern prairie skinks in Louisiana, but in other parts of their geographic range they are active at dusk and dawn as well as during the day. They move about on the ground in search of prey, and when threatened typically retreat beneath vegetation or into a hole in the ground.

FOOD AND FEEDING Prairie skinks eat insects and other small invertebrates.

The prairie skink is found in northwestern Louisiana but is primarily a western species.

REPRODUCTION Mating occurs in the spring, and females lay 6–10 eggs in May and June and guard them until they hatch after about 5 or 6 weeks.

PREDATORS AND DEFENSE Snakes and predatory birds are probably the most common predators of southern prairie skinks.

CONSERVATION ISSUES Conservation of this species is not an issue in the Southeast because of its marginal occurrence.

The Florida reef gecko is the only gecko native to the Southeast.

Florida Reef Gecko   *Sphaerodactylus notatus*

FAMILY Gekkonidae

DESCRIPTION Florida reef geckos are the tiniest lizards native to the United States and among the smallest lizards in the world. They have a distinctive flat, pointed snout and a pair of lidless, unblinking brown eyes. The body is generally brown. Females and juveniles have brown and yellowish striping on the head; males have dark spots and speckles all over. The scales on the back are relatively large compared with those of other small geckos and have distinct longitudinal keels.

VARIATION AND TAXONOMIC ISSUES Of the 13 species of geckos known from Florida by the end of the twentieth century, the reef gecko is the only one native to the United States. Besides the Florida subspecies (*S. notatus notatus*), four other subspecies occur on some Caribbean islands, including Cuba and the Bahamas.

WHAT DO THE HATCHLINGS LOOK LIKE? Hatchlings have two dark-bordered light spots on the back of the neck and look like smaller versions of the females.

BODY PATTERN AND COLOR
body brownish, sometimes with dark spotting

DISTINCTIVE CHARACTERS
pointed snout; vertical pupils; no eyelids; overlapping keeled scales on back

SIZE

2.25"   1"

● ADULT
● HATCHLING

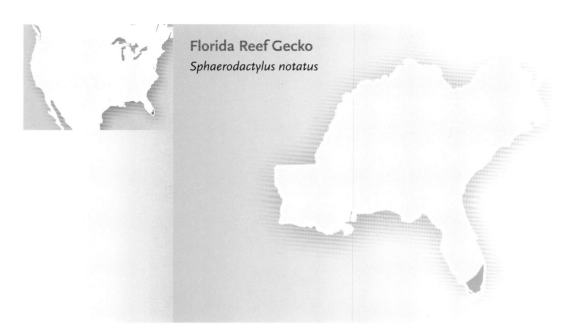

Florida Reef Gecko
Sphaerodactylus notatus

The flat, pointed snout and lidless eyes are characteristic of the Florida reef gecko.

CONFUSING SPECIES This tiny lizard is most likely to be confused with the ocellated and ashy geckos that have been introduced into the Florida Keys and extreme southern peninsular Florida. Both of the introduced forms have light markings on the head and body, and the ashy gecko has smaller, smooth, nonoverlapping scales on the back.

DISTRIBUTION AND HABITAT Florida reef geckos are found from Dry Tortugas National Park and Key West throughout the Florida Keys and onto the Atlantic side of mainland Florida into Miami-Dade and Broward counties. They occur around man-made structures and in cleared lots and gardens, as well as in natural pine habitat, on and around rocks near beaches, and in hammocks (vegetated habitats that are slightly elevated from the surrounding area).

BEHAVIOR AND ACTIVITY These little lizards are common but very secretive. They hide beneath leaf litter and debris, and run for cover if their hiding place is disturbed. They are most active on the surface in early evening and at night. Individuals have been observed licking their lips and their eyes.

FOOD AND FEEDING Florida reef geckos primarily eat small insects, with ants, flies, and springtails probably being common prey items.

REPRODUCTION Females lay a single, tiny (less than 0.25 inch long), oval egg at a time. Eggs are laid year-round beneath ground cover or in dead pine needles in crotches of trees, sometimes more than 6 feet above the ground.

PREDATORS AND DEFENSE These lizards are probably preyed on by larger lizards and snakes.

CONSERVATION ISSUES A primary concern in the conservation of this species is competition from introduced species of lizards, particularly other geckos that may be predators or competitors.

Tiny Florida reef geckos can be very difficult to spot, sprinting for cover the moment their hiding place is discovered.

Did you know?

The venomous Gila monster (Heloderma suspectum) almost never bites unless it is picked up, and although the bite is excruciatingly painful, deaths from the venom are very rare.

INTRODUCED LIZARDS

Collared lizards found in Louisiana are considered to be introductions from their native southwestern geographic range.

Collared Lizard

Crotaphytus collaris

FAMILY Iguanidae

DESCRIPTION These heavy-bodied, big-headed lizards have a distinct black-and-white collar across their shoulders. Females have a brownish body; males are more colorful, with a mostly greenish body and brownish head. The largest males are more than a foot long.

VARIATION AND TAXONOMIC ISSUES Some herpetologists place this species in the family Crotaphytidae.

ORIGIN AND DISTRIBUTION Individuals of this species, native to the U.S. Southwest, have been reported from two parishes in western Louisiana along the Texas border. The reports may have been based on observations of accidental hitchhikers, however. No established population of collared lizards has been confirmed in Louisiana.

The distinctive black collar of male collared lizards is more vivid than that of females.

How do you identify a collared lizard?

BODY PATTERN AND COLOR
color variable; males usually with greenish body, brown head; females brown; both sexes with black-and-white neck band

DISTINCTIVE CHARACTERS
large head; run upright on hind legs

DISTRIBUTION IN THE SOUTHEAST
western Louisiana

SIZE

12"

Adult collared lizards consume many insects but will also eat other lizards as well as small mammals.

NATURAL HISTORY Collared lizards are typically found in arid habitats, often in rocky terrain. They are voracious predators of invertebrates, reptiles—including smaller lizards—and small mammals. Among their notable behavioral traits is their ability to run upright on their hind legs.

CONSERVATION ISSUES No conservation concerns are warranted for collared lizards in the Southeast unless populations become established in areas outside their natural range.

An introduced Texas horned lizard, with its characteristic head and body spines, is unlikely to be confused with any other lizard species found in the Southeast.

Texas Horned Lizard   *Phrynosoma cornutum*

FAMILY Iguanidae

DESCRIPTION Texas horned lizards are unlikely to be mistaken for any other species of southeastern lizard. The broad, brownish, flattened body is covered with spines, and the head bears two larger spines, or "horns." Adults are usually 3–4 inches long and have a very short tail.

VARIATION AND TAXONOMIC ISSUES Although several species of horned lizards are native to the western United States, the Texas horned lizard—also known as the horned toad—is the only one known to have become established in the Southeast. Some herpetologists place this species in the family Phrynosomatidae.

Baby horned lizards look like miniature adults.

BODY PATTERN AND COLOR brownish with dark markings

DISTINCTIVE CHARACTERS flat body; short tail; spiny appearance with row of spines around outside of body

DISTRIBUTION IN THE SOUTHEAST isolated populations in North Carolina, South Carolina, and Florida

SIZE

4"

ORIGIN AND DISTRIBUTION The species is native from Kansas through Texas to northern Mexico. Texas horned lizards have been observed in several parishes in Louisiana, a dozen counties in Florida, and scattered sites in Alabama, Georgia, and the Carolinas. No established populations are known in Louisiana, but there are at least two in northern Florida—one in Duval County near Jacksonville and the other in Escambia County on Santa Rosa Island. Two breeding populations are known from South Carolina coastal areas, one in Charleston County at Isle of Palms and one in Colleton County at Edisto Beach. A small but persistent population lives in coastal Onslow County, North Carolina. All introduced populations presumably originated from pets released in suitable habitat.

NATURAL HISTORY Texas horned lizards occupy dry, sandy habitats characteristic of coastal areas and are found in sunny areas rather than shade. They eat native ants as well as a variety of other small insects and centipedes. Horned lizards have a peculiar antipredator defense, especially when threatened by a dog or coyote: they squirt blood from their eyes. Whether southeastern populations use this defense is unknown. They defend themselves against imported fire ants in two ways: if only a few ants are present, the horned lizard will remain still and eat any ants that run across its mouth; if many ants are present, such as at a disturbed ant mound, the lizard will run several feet away and burrow backward into the sand.

CONSERVATION ISSUES Ironically, while some of the introduced populations in the Southeast have persisted for decades, the Texas horned lizard can no longer be found at many sites in Texas where it was once common.

Hispaniolan green anoles climb up trees and onto low-lying vegetation.

Hispaniolan Green Anole *Anolis chlorocyanus*

FAMILY Iguanidae

DESCRIPTION These medium-sized lizards resemble our native green anole, and like them can change color from green to brown. The dewlap is usually bluish gray.

VARIATION AND TAXONOMIC ISSUES The Hispaniolan green anole is also known as the Haitian green anole or blue-green anole. Two subspecies have been described in their natural range.

ORIGIN AND DISTRIBUTION As the common name indicates, the Hispaniolan green anole originated in Hispaniola. The small south Florida populations are the only ones known to be present in the United States.

NATURAL HISTORY Hispaniolan green anoles live high in trees or buildings and are seldom seen less than 6 feet above the ground. Little else is known about their ecology in Florida.

CONSERVATION ISSUES The species may be unable to persist in areas where large trees are removed.

The Hispaniolan green anole is a medium-sized green or brown lizard with a dewlap that is usually bluish gray.

7"

Crested Anole

Anolis cristatellus

FAMILY Iguanidae

DESCRIPTION This medium-sized lizard (up to 7 inches in total length) has a thick, brownish body that sometimes bears darker brown markings or stripes. The dewlap is commonly orangish. A very distinct crest runs down the middle of the back from the head onto the tail.

VARIATION AND TAXONOMIC ISSUES The crested anole is also called the Puerto Rican crested anole. Two subspecies have been described; *A. c. cristatellus* is the one introduced into Florida.

The crested anole's dewlap has an orange tint.

ORIGIN AND DISTRIBUTION Crested anoles, which originated in Puerto Rico and the Virgin Islands, have been known from a few small populations in Miami-Dade County, Florida, since the 1970s. The species was reported from Brevard County in the 1990s, but later visits to the same site turned up no specimens.

NATURAL HISTORY Crested anoles are tree dwellers that also will occupy the sides of buildings, fence posts, hedges, walls, and other man-made structures, although juveniles may be active on the ground near trees. They eat a variety of insects, and smaller individuals also nibble on flowers and small fruit.

CONSERVATION ISSUES The species does not appear to be a major threat to any native lizard.

Crested anoles prefer to dwell in trees or up above the ground on other raised structures, such as buildings or fences.

Large-headed anoles are often found on low tree branches.

How do you identify a large-headed anole?

BODY PATTERN AND COLOR
light to dark brown; faint stripes down sides

DISTINCTIVE CHARACTERS
pale yellow dewlap

DISTRIBUTION IN THE SOUTHEAST
Miami-Dade, Broward, and Martin counties, Florida

SIZE

8"

Large-headed Anole

Anolis cybotes

FAMILY Iguanidae

DESCRIPTION These medium-sized brown lizards (up to 8 inches in total length) are not capable of changing to green. The brown can range from light to dark in the same individual depending on temperature or mood. The body has light stripes down the sides and darker markings on the back. The dewlap is pale yellow.

VARIATION AND TAXONOMIC ISSUES Several subspecies of large-headed anoles have been described; the one occurring in Florida is generally referred to as *A. c. cybotes*. Some authorities place this species in the genus *Ctenonotus*.

The dewlap of a large-headed anole is pale yellow.

ORIGIN AND DISTRIBUTION Large-headed anoles originated in Hispaniola and the Bahamas. At least three established populations are present in the extreme southeastern tip of Florida. A colony has been in Miami-Dade County since the early 1970s, and persistent populations have been reported from Broward and Martin counties.

NATURAL HISTORY Large-headed anoles can be found on the ground as well as on the lower trunks of trees or the sides of buildings, fences, and other man-made structures. Insects are probably their primary prey, although they are known to eat smaller lizards.

CONSERVATION ISSUES The species appears to remain in localized populations in disturbed habitat and is unlikely to compete with native lizards.

True to their name, large-headed anoles have disproportionately large heads.

The dull-colored bark anole appears to be maintaining populations and increasing its range in Florida.

How do you identify a bark anole?

BODY PATTERN AND COLOR
green, brown, or gray; sometimes mottled

DISTINCTIVE CHARACTERS
pale yellow dewlap, often with orange spot; short snout and body

DISTRIBUTION IN THE SOUTHEAST
southern Florida from the Lower Keys to Lake Okeechobee and Fort Myers

SIZE

5"

Bark Anole

Anolis distichus

FAMILY Iguanidae

DESCRIPTION Bark anoles are smaller than most of the other introduced anoles (maximum length about 5 inches) and have a relatively shorter nose. The body color is highly variable but can be brownish, drab green, or gray, and is often mottled gray and brown, a coloration that is ideal camouflage on most tree trunks. The dewlap is also variable in the Florida populations but is typically pale yellowish to almost white, sometimes with an orange spot in the center.

The bark anole's dewlap may be whitish.

VARIATION AND TAXONOMIC ISSUES This species is separated into 18 subspecies, of which at least 5 have been reported from Florida. *Anolis distichus floridanus* was actually described on the basis of specimens from Miami and is considered by some lizard biologists to be a native U.S. lizard. The widespread occurrence of adjacent populations of the species in many parts of Florida has led to interbreeding that makes it difficult to determine the subspecies to which a particular individual belongs. Different populations of these lizards are called Florida bark anoles and green bark anoles. Some authorities place this species in the genus *Ctenonotus*.

ORIGIN AND DISTRIBUTION The species originated in Hispaniola and the Bahamas. Populations occur from the lower Keys northward in the Atlantic Coast counties as far north as Lake Okeechobee, and in Lee County on the Gulf side. Populations appear to be persisting in many areas, and the species has expanded its geographic range since the earliest introductions.

NATURAL HISTORY Like many of the other introduced anoles, bark anoles have an affinity for large, smooth-trunked trees such as ficus and banyan, as well as buildings and other man-made structures. Individuals characteristically stay 3–6 feet above the ground on tree trunks. They use speed and agility to race away from predators. These little anoles eat insects and apparently are especially fond of ants and small beetles. Because of their small size they fall prey to larger lizards such as the Cuban green anole.

CONSERVATION ISSUES The species is not known to compete with native lizards.

Bark anoles prefer to stay 3–6 feet above the ground on tree trunks or manmade structures.

Knight anoles are the largest of the several species of introduced anoles in Florida. The species was first known to be present in persistent populations in Florida in the 1950s.

How do you identify a knight anole?

BODY PATTERN AND COLOR
usually green, occasionally brown; yellow jaw stripes below eye and on shoulder; juveniles green with light banding on back and tail

DISTINCTIVE CHARACTERS
large, pink dewlap; head large in proportion to body

DISTRIBUTION IN THE SOUTHEAST
Miami-Dade, Monroe, Broward, Collier, and possibly other counties in Florida

SIZE

18"

Knight Anole

Anolis equestris

FAMILY Iguanidae

DESCRIPTION Knight anoles are the largest anoles in Florida and may reach a total length of more than 18 inches. They are typically green but can change color to light or dark brown. Two distinctive features are the presence of a yellowish stripe on the upper jaw below the eye and another stripe slanting upward from the lower jaw and across the shoulder. Both sexes have large dewlaps that are light pink when extended. Juvenile knight anoles are green and have thin, obscure yellowish bands across the back and tail in addition to the yellowish lip stripe of adults.

VARIATION AND TAXONOMIC ISSUES Several subspecies have been described in Cuba, but the one prevalent in Florida is *A. e. equestris*. Other common names include Cuban knight anole and chipojo, and locals sometimes refer to them erroneously as iguanas.

ORIGIN AND DISTRIBUTION Knight anoles, which originated in Cuba, were first reported to be established in Florida in the 1950s. They have been seen in the four southernmost counties in Florida and in a few other counties in the lower half of the Florida peninsula. Large populations are present throughout much of Miami-Dade County. The species is also established in Hawaii.

A juvenile knight anole in southern Florida perches on a tree branch.

NATURAL HISTORY Knight anoles are among the better-studied introduced lizards because of their large size, persistence in observable populations in Florida for more than half a century, and conspicuous presence in many urban areas. They are generally associated with large trees, especially black olive and ficus. Knight anoles are less active from November through March when temperatures are below about 80°F (27°C), remaining in the canopy of tall trees. From April through August they come out of the canopy and closer to the ground, often perching on tree trunks (head down) only 5 or 6 feet above the ground. The diet generally consists of large insects, including butterflies and beetles, but knight anoles will also eat spiders, Cuban treefrogs, smaller lizards, small birds, bird eggs, and a variety of fruits, including figs, palm fruits, and ripe mangos. Knight anoles can live for several years.

The knight anole's light pinkish dewlap and yellow jaw stripes distinguish the species from others.

CONSERVATION ISSUES This species could reduce the abundance of some native species, including lizards, by predation, but their impact is most likely to occur in areas that have already been modified through commercial development and landscaping. They are commonly collected in Florida neighborhoods for the pet trade.

Jamaican giant anoles can range in color from brown to green.

Yellow coloration is apparent in the dewlap of a Jamaican giant anole.

How do you identify a Jamaican giant anole?

BODY PATTERN AND COLOR striking green body, or sometimes brownish

DISTINCTIVE CHARACTERS dewlap yellowish; low but apparent jagged crest down center of body and tail

DISTRIBUTION IN THE SOUTHEAST Miami-Dade and possibly Lee and Martin counties, Florida

SIZE

10"

Jamaican Giant Anole

Anolis garmani

FAMILY Iguanidae

DESCRIPTION These large, beautiful lizards (males may be more than 10 inches long) range in basic body color from brown to vivid green with a yellowish dewlap. A slightly elevated crest runs down the center of the body and tail.

VARIATION AND TAXONOMIC ISSUES No subspecies are recognized; some authorities place the species in the genus *Norops*.

ORIGIN AND DISTRIBUTION Jamaican giant anoles, which are indeed from Jamaica, were established by the mid-1970s in Miami-Dade County, Florida. By the mid-1980s they were present in two other counties (Lee and Martin), but those populations disappeared during the unusually cold winter of 1991.

NATURAL HISTORY Jamaican giant anoles live in large trees, often climbing more than 30 feet up but also descending to 2 or 3 feet above the ground. They eat most species of insects they can capture, other lizards, and ripe fruit.

CONSERVATION ISSUES The species probably has little impact on any native lizards or other species because of its localized range.

Some lizard biologists believe the Cuban green anole has hybridized with the native green anole in some areas of Florida. Others believe that the two may belong to the same species.

How do you identify a Cuban green anole?

BODY PATTERN AND COLOR
green or brown; individuals can change color

DISTINCTIVE CHARACTERS
dewlap pink or reddish; virtually indistinguishable from the native green anole; long, skinny tail

DISTRIBUTION IN THE SOUTHEAST
Miami-Dade County, Florida

SIZE

7"

Cuban Green Anole

Anolis porcatus

FAMILY Iguanidae

DESCRIPTION These green lizards look almost exactly like green anoles, and like the native species are capable of turning brown; they also have a pink or reddish dewlap. They are so similar in appearance and behavior to the green anole that the two are difficult to tell apart.

VARIATION AND TAXONOMIC ISSUES This lizard is also referred to as *chameleo verde*. The species is suspected of hybridizing in southern Florida with the green anole, and some authorities even consider them the same species. Two subspecies of *A. porcatus* have been described in Cuba.

ORIGIN AND DISTRIBUTION The first specimens of this Cuban species were collected in 1937 on Key West. Since then, Cuban green anoles have been

The Cuban green anole's dewlap is pink or reddish.

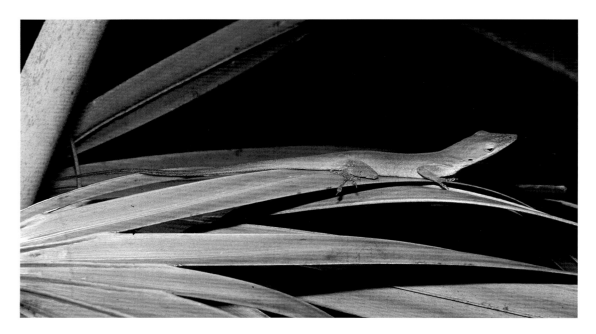

Cuban green anoles have been witnessed licking ornamental palm trees for the nectar and eating the palms' flowers.

found in only a few isolated localities in Miami-Dade County. They may be more widespread in southern Florida than is apparent, however, and mistaken for the native green anoles.

NATURAL HISTORY Cuban green anoles look and behave like green anoles and as far as is known have the same basic ecology. Although primarily insect eaters, Cuban green anoles in Florida have been observed licking nectar from and eating the flowers of an ornamental palm tree. This species also eats smaller anoles, such as the bark anole.

CONSERVATION ISSUES The greatest potential impact of this species on native fauna is its likelihood of interbreeding with the green anole.

The first U.S. specimens of the introduced Cuban green anole were found in Key West, Florida, in the 1930s.

How do you identify a brown anole?

BODY PATTERN AND COLOR gray to dark brown; wavy white line down back

DISTINCTIVE CHARACTERS individuals can change to varying shades of brown; dewlap reddish orange, bordered by white

DISTRIBUTION IN THE SOUTHEAST peninsular Florida and most of the Panhandle; isolated populations in Georgia, Alabama, Louisiana, and possibly South Carolina

SIZE

8"

Brown Anole   *Anolis sagrei*

FAMILY Iguanidae

DESCRIPTION Brown anoles are small to medium-sized lizards (up to about 8 inches including the tail) that can change color from gray to brown to almost black, but are never green. Reticulated markings are sometimes apparent on the body, and a wavy pattern and white stripe may be visible down the center of the back. The dewlap is typically more orange than red with light coloration around the perimeter. Males may exhibit a ridge along the neck and partially down the back.

VARIATION AND TAXONOMIC ISSUES Other common names include Cuban anole, Cuban brown anole, Bahamian brown anole, and Stejneger's anole. Six subspecies have been described, and two of them—*A. s. sagrei* from Cuba and *A. s. ordinatus* from the Bahamas—have been introduced into

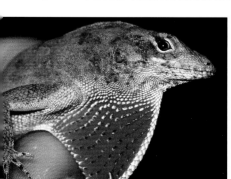

Lighter coloration around the edge of the reddish orange dewlap is characteristic of the brown anole.

Brown anoles can vary in body color from brown to gray to black but are never green.

Florida. A third subspecies, *A. s. stejnegeri*, was initially described as a separate species in 1931 from specimens captured in Key West, Florida, but was later classified as a subspecies. Some authorities place *A. sagrei* along with the Jamaican giant anole in the genus *Norops*, which is closely related to *Anolis*.

ORIGIN AND DISTRIBUTION This species, which originated in the West Indies, is more widely distributed in the Southeast than any other introduced anole and has apparently been here since the late 1800s. Populations have now been reported from all of the counties in peninsular Florida, most of those in the Florida Panhandle, and several counties in southern and central Georgia. The species is also established in parts of southern Alabama and several parishes in Louisiana, and isolated sightings have been made in coastal and central South Carolina.

Although the species was intentionally introduced in some locations, individual brown anoles have probably been carried unintentionally as hitchhikers in shipments of plants and garden supplies, lumber, and building products to every southeastern state and some midwestern and northern ones. Single individuals and small populations were more often observed at motels and interstate rest stops than elsewhere in the 1990s and early 2000s, suggesting transport by vehicles. Warm winter conditions appear to be necessary for a population to become established, but gradual evolutionary adjustment to colder climatic conditions could result in brown anoles continuing to move northward. The species has been reported from some areas of southern Texas and is a successful exotic in Hawaii.

NATURAL HISTORY Brown anoles are easily observed in the many developed areas where dense populations occur (densities of one lizard per square yard have been reported). They seem to be everywhere: climbing or basking on trees or other vegetation; on the sides of buildings, fences, or other man-made structures; and darting about on the ground from the base of one plant to another. Because they are so abundant in some locations, we know more about certain aspects of brown anole ecology and behavior than we do about most other introduced lizards. In the Miami area, brown anoles reproduce continually from April until September, with females laying a single egg beneath vegetation or in soft soil about once a week. The incubation period is about 4 weeks, and the hatchlings are slightly more than an inch long.

Small insects and spiders constitute most of the diet, but brown anoles have often been observed eating smaller green anoles as well as cannibalizing individuals of their own species. Documented predators of adult brown anoles, their eggs, and hatchlings include native green anoles, several of the larger introduced lizards, Cuban treefrogs, ringneck and corn snakes, predatory birds (including great egrets), and spiders.

The brown anole is one of the most widespread nonnative lizards in the Southeast.

Research in the Bahamas has shown that these superabundant lizards can dramatically influence species diversity, community composition, and entire food webs by being both predators and prey.

CONSERVATION ISSUES The brown anole and the native green anole eat similar insect prey and may thus compete for resources, and adults of both species will eat small individuals of the other. Green anoles shift their habitat use upward and into more heavily vegetated areas when brown anoles are present. A long-lasting impact on the distribution and abundance of green anoles remains to be confirmed, but the brown anole's expanding geographic range in the Southeast, high abundance, and significant effect on some food webs suggest that the species may have notable environmental impacts on green anoles, on other native species of lizards, and potentially on other fauna.

Brown anole males may exhibit a ridge down their necks and part way down their backs.

An adult male green iguana from Broward County, Florida, in full breeding colors

How do you identify a green iguana?

BODY PATTERN AND COLOR
body color mixed green, brown, and gray; males sometimes with bright orange head and front of body; tail with black bands; babies green

DISTINCTIVE CHARACTERS
center of back and tail ornamented with spines

DISTRIBUTION IN THE SOUTHEAST
established populations in Miami and the Florida Keys; isolated sightings may occur anywhere in the Southeast in warmer months

SIZE

6'

Green Iguana

Iguana iguana

FAMILY Iguanidae

DESCRIPTION Green iguanas are the largest of the lizards native to the Western Hemisphere that have breeding populations in the United States. Individuals commonly reach a total length of 5 feet, and many are more than 6 feet long. The body and legs are shades of green, brown, and gray, and the tail is typically encircled by black bands. Adult males have more ornamentation than females, but both have a frill of spines down the center of the back and tail. Mature males have a large, permanent dewlap and often develop bright orange coloration on their neck and front legs. Juvenile iguanas are bright green.

VARIATION AND TAXONOMIC ISSUES Green iguanas are often called simply iguanas, or sometimes bamboo chickens. Subspecies have been described but are generally not considered valid.

ORIGIN AND DISTRIBUTION The species occurs naturally from Mexico south throughout Central and South America and on many islands of the Caribbean. Introduced green iguanas have been reported from several counties in Florida, including as far north as Alachua County (Gainesville), as well as in urban areas of other southeastern states. Most of those sighted outside south Florida were probably released or escaped pets that did not survive the winter; however, breeding populations have been known in Miami-Dade County, Florida, since at least the 1970s, and green iguanas were first sighted at least a decade before that. Several established populations exist in developed areas around Miami, in the Florida Keys, and possibly farther north. Green iguanas have also been introduced into Hawaii.

NATURAL HISTORY Green iguanas generally inhabit large trees that are near water. When threatened, they escape to the canopy or plunge into a canal or other body of water and swim to the bottom or the other side to escape. They are mostly herbivorous, eating flowers, fruits, and leaves, but they will also eat insects and other small animals. Young iguanas are most likely to eat insects. Native predators include yellow-crowned night-herons, Florida burrowing owls, hawks, gray foxes, and raccoons. Removal of raccoons from state parks by the Florida Park Service has resulted in noticeable increases in iguana numbers, suggesting that these mammalian predators control population sizes of green iguanas. Green iguanas are capable of living for many years, as evidenced by a record of a captive female that was kept as a pet for almost 28 years.

CONSERVATION ISSUES Their large size and bright coloration make green iguanas noticeable additions to the local fauna, but because of their herbivorous habits, their impact on native lizards is probably minimal.

Male juvenile green iguanas do not have well-developed ornamentation.

Green iguanas bask on a limb above a canal in Key Biscayne, Florida.

Mexican spinytail iguanas may reach lengths of four feet.

How do you identify a Mexican spinytail iguana?

BODY PATTERN AND COLOR
dark body and tail with darker bands; juveniles green with black bands

DISTINCTIVE CHARACTERS
spines down center of back; tail with spiny appearance; not always readily distinguishable from black spinytail iguana, but darker bands occur more toward head and neck

DISTRIBUTION IN THE SOUTHEAST
Miami-Dade County, Florida

SIZE

4'

Mexican Spinytail Iguana   *Ctenosaura pectinata*

FAMILY Iguanidae

DESCRIPTION These lizards change in color from bright green as juveniles to nondescript but usually dark adults with a black head and black banding on the body. A row of elevated spines runs down the back, and a series of rings of scales give the tail a spiny appearance. The largest adults reach 4 feet in total length and look very much like black spinytails.

VARIATION AND TAXONOMIC ISSUES Some introduced populations in Florida that were originally identified as *C. pectinata* were later shown to be the more common black spinytail iguana (*Ctenosaura similis*).

ORIGIN AND DISTRIBUTION Although incidental reports of this Mexican species are common in Florida because of accidental or intentional releases of pets, breeding populations are uncommon. An established population is in the community of Perrine in Miami-Dade County.

NATURAL HISTORY Mexican spinytail iguanas are associated with rock piles, rock walls, and abandoned man-made structures where they can hide and dig their burrows. Although mostly herbivorous, they will also eat insects and small vertebrates.

CONSERVATION ISSUES Because they are herbivorous and highly localized, Mexican spinytail iguanas probably do not compete with native lizards or pose a threat to them.

Male black spinytail iguanas have dark crossbands on the body, a row of spines down the center of the back, and whorls of enlarged, spiny scales on the tails.

Black Spinytail Iguana *Ctenosaura similis*

FAMILY Iguanidae

DESCRIPTION These large lizards commonly reach 3 feet in total length, and some reach 4 feet. Body coloration of adults is highly variable, ranging from drab light or dark gray to almost black but including browns, yellows, and even blue or green shading; four to six dark crossbands are often apparent on the body. The dewlap is visible, and a row of spines runs down the center of the back from the head to the tail. The tail itself is ringed with whorls of enlarged spiny scales and has a crest of scales down the center. Young spinytails are green.

VARIATION AND TAXONOMIC ISSUES Other common names include black iguana and Central American spinytail iguana.

Juvenile black spinytail iguanas are green.

How do you identify a black spinytail iguana?

BODY PATTERN AND COLOR body light to dark gray mottled with brown, green, yellow, and blue, with dark bands

DISTINCTIVE CHARACTERS spines down center of back; tail with spiny appearance; not always readily distinguishable from Mexican spinytail iguana, but spines more pronounced and dark banding more toward mid-body than toward head

DISTRIBUTION IN THE SOUTHEAST scattered localities in southern Florida, including coastal islands in Collier County

SIZE

3'

Black Spinytail Iguana • 143

ORIGIN AND DISTRIBUTION The black spinytail iguana originated in Mexico and Central America. Populations are reportedly established at numerous locations in coastal southern Florida as far north as Tampa on the Gulf Coast and Broward County on the Atlantic Coast. Large, persistent colonies are present on Gasparilla Island (Charlotte County) and Keewaydin Island (Collier County).

NATURAL HISTORY Black spinytail iguanas prefer rock outcrops, broken concrete, and rock walls that provide crevices for hiding and surfaces on which to bask. Individuals typically retreat when approached, but larger ones may actually hold their ground with threat displays. They are omnivorous, eating fruits, seeds, and leaves, but also preying on insects and a variety of small vertebrates. Females dig nests in soft soil, deposit their eggs, and then cover them.

CONSERVATION ISSUES Black spinytail iguanas have been suggested as potential predators of native fauna, including baby sea turtles and nestling birds, in some areas. They also eat natural and ornamental vegetation used in landscaping and are considered pests for that reason.

Did you know?

Marine iguanas of the Galápagos Islands dive more than 30 feet beneath the ocean's surface to feed on algae growing on rocks.

The black spinytail iguana will look for a habitat with both crevices for hiding and surfaces for basking.

Brown Basilisk

Basiliscus vittatus

FAMILY Iguanidae

DESCRIPTION Brown basilisks are narrow-bodied brown lizards with noticeably longer legs, especially the rear ones, than other southeastern lizards. Their basic body color is brown, with smaller individuals having yellow or white stripes across the face and down the sides. Some individuals may have crossbanding on the body. Adult males have a conspicuous crest behind the head and down the body and can reach a total length exceeding 2 feet.

VARIATION AND TAXONOMIC ISSUES This species has also been called the striped basilisk or Jesus Christ lizard. Some authorities place the species in the family Corytophanidae.

Juvenile brown basilisks have yellowish stripes on the face and sides.

ORIGIN AND DISTRIBUTION The brown basilisk, which originated in Mexico, Central America, and northern South America, has successfully colonized and dispersed from several localities along the Atlantic side of southern Florida in Miami-Dade, Broward, and Palm Beach counties. The species was first sighted in Florida in the 1970s.

NATURAL HISTORY A noted feature of basilisks is their association with aquatic habitats. In southern Florida they are particularly prevalent in thick vegetation and trees alongside open canals. They often sleep in trees or vegetation but bask openly on pavement or the ground. When threatened, they may retreat into heavy vegetation or up a tree, but many will head for the water, where they dive below the surface or, in a spectacular display, run rapidly across the top of the water to the other side. Brown basilisks are primarily insect eaters but also eat fruit and small vertebrates.

CONSERVATION ISSUES The brown basilisk is not likely to have an impact on native lizards.

Did you know?

Some lizards, such as the tropical American basilisks, can run upright across the surface of open water.

While brown basilisks may retire in thick vegetation or trees, they openly bask on exposed ground or pavement.

Northern curly-tailed lizards are thick-bodied and reach total lengths of up to 10 inches. They will eat brown anoles and possibly other lizards.

How do you identify a northern curly-tailed lizard?

Northern Curly-tailed Lizard *Leiocephalus carinatus*

FAMILY Iguanidae

DESCRIPTION These chunky, medium-sized lizards (average total length about 8 inches, up to 10 inches) are gray, light brown, or dark brown above, lighter on the lower chin and belly, and have a short stubby nose. Pointed, keeled scales that form a low ridge down the center of the back give the skin a rough texture.

VARIATION AND TAXONOMIC ISSUES Lion lizard is another common name. Several subspecies are known from various islands. The one initially introduced into Florida and believed to be the prevalent form in the state is *L. c. armouri* from the Bahamas. At least two other subspecies (*L. c. virescens* and *L. c. coryi*) have been introduced into Florida, although there is no indication that either has become established. We have included the genus *Leiocephalus* in the all-inclusive family Iguanidae, but some taxonomists place the genus in a separate family, Tropiduridae, and others put it in the family Leiocephalidae.

ORIGIN AND DISTRIBUTION Northern curly-tailed lizards originated on the Caribbean islands. The subspecies *L. c. virescens* was introduced into Florida in the 1930s. By the early 2000s, populations were established in

BODY PATTERN AND COLOR
body drab gray or brown; chin whitish

DISTINCTIVE CHARACTERS
tail of adult males held in tight curl above back during spring breeding season; keel down back

DISTRIBUTION IN THE SOUTHEAST
Monroe, Miami-Dade, Broward, Palm Beach, Martin, Indian River, and Brevard counties, Florida

SIZE

10"

Northern curly-tailed lizards often retreat to underground burrows.

numerous localities in Atlantic Coast counties from Miami-Dade County to Brevard County (Cocoa Beach) and as far south as Key West.

NATURAL HISTORY Northern curly-tailed lizards occupy sandy or rocky habitat in both natural areas and disturbed areas with man-made structures. They retreat to small underground burrows when threatened. Insects are the primary food, but these lizards are also said to eat small brown anoles. Little blue herons and mockingbirds are known predators. Adult males curl the tail above the back during the mating season.

CONSERVATION ISSUES The northern curly-tailed lizard is reported to eat brown anoles and could possibly have a localized effect as a predator on native green anoles.

Populations of the red-sided curly-tailed lizard are known from several locations in Miami-Dade and Broward counties in Florida.

How do you identify a red-sided curly-tailed lizard?

Red-sided Curly-tailed Lizard

Leiocephalus schreibersi

FAMILY Iguanidae

DESCRIPTION Red-sided curly-tailed lizards are medium-sized, thick-bodied lizards (average total length about 8 inches, up to 10 inches) with a small head and short stubby nose. The body is typically dull gray or brownish, with the sides brightened by reddish coloration and small lighter-colored spots. A fold of skin with smaller scales than those on the back runs the length of the body below the midline. Keeled scales form a low ridge down the center of the back.

VARIATION AND TAXONOMIC ISSUES Also called Hispaniolan curly-tailed lizard. Two subspecies have been described. Some authorities place the genus in the family Tropiduridae.

ORIGIN AND DISTRIBUTION Although not as widespread as the northern curly-tailed lizard, red-sided curly-tailed lizards, which originated in Hispaniola, occur in a few small, localized, but well-established populations in Miami-Dade and Broward counties.

BODY PATTERN AND COLOR
body gray or brown; sides reddish with whitish spots

DISTINCTIVE CHARACTERS
fold of skin on sides; scales down center of back form a keel

DISTRIBUTION IN THE SOUTHEAST
Miami-Dade and Broward counties, Florida

SIZE

10"

Keeled scales form a low ridge down the back of the red-sided curly-tailed lizard.

NATURAL HISTORY Little is known of this species' habits in Florida aside from its occurrence in open disturbed habitats with areas for hiding. They eat mostly insects. Males curl the tail over the back during the mating season, but not as tightly as northern curly-tailed lizards do.

CONSERVATION ISSUES The red-sided curly-tailed lizard has no known competitive or predatory interactions with native lizards.

A large adult male rainbow whiptail is one of Florida's most colorful lizards.

How do you identify a rainbow whiptail?

Rainbow Whiptail

Cnemidophorus lemniscatus

FAMILY Teiidae

DESCRIPTION The rainbow whiptail is a relatively large, streamlined lizard (total length may exceed a foot) with an exceptionally long tail. The males are spectacularly colored in blues, yellows, and greens, including a blue face and throat. Females have an orange head. Both sexes usually have stripes that run the length of the body and onto the tail.

VARIATION AND TAXONOMIC ISSUES Lizard biologists are uncertain where the introduced animals originated, and the forms that have become established may differ from each other biologically. Rainbow whiptails are sometimes called rainbow lizards.

ORIGIN AND DISTRIBUTION This tropical American species occurs in small populations in Miami-Dade County, Florida. The earliest records in Florida are from the 1960s.

NATURAL HISTORY Rainbow whiptails are active during the day and usually remain on the ground rather than climbing trees. Typical habitat includes vegetated areas with numerous open spaces alongside railroad tracks or other disturbed habitats. When threatened, they rely on their speed to outrun predators and retreat to underground burrows. Females lay eggs,

BODY PATTERN AND COLOR males colorful, with blue face and blue, yellow, and green body stripes; brown stripe down center of back; head of female orange; both sexes may have stripes down the body

DISTINCTIVE CHARACTERS streamlined appearance; long tail

DISTRIBUTION IN THE SOUTHEAST Miami-Dade County, Florida

SIZE

12"

Introduced rainbow whip-
tails were first reported
in the Miami area in the
1960s.

but whether or not any reproduce parthenogenetically (i.e., produce viable eggs that are not fertilized by a male) as some forms of the species do is unknown. Males are present in some of the introduced populations. Rainbow whiptails eat predominantly insects but will also eat vegetation. The corn snake is a known predator, and other lizard-eating snakes probably eat these lizards too.

CONSERVATION ISSUES The range of the rainbow whiptail lies within that of the six-lined racerunner, and the two species may compete under certain circumstances.

Introduced giant whiptails reach lengths of a foot or more and are known to occur in localities in the Miami area.

Giant Whiptail

Aspidoscelis motaguae

FAMILY Teiidae

DESCRIPTION This is the largest of the introduced *Aspidoscelis* species. Adults may be more than a foot in total length. The brownish body and grayish sides have numerous light yellowish to white spots. The underside is blue, and the tail is reddish brown.

VARIATION AND TAXONOMIC ISSUES This species, sometimes called the Central American whiptail, was formerly placed in the genus *Cnemidophorus*.

ORIGIN AND DISTRIBUTION Giant whiptails originated in Central America and have been reported from two localities in Miami-Dade County, Florida.

NATURAL HISTORY These lizards are most common in sandy terrain with sunny, open areas and sparse vegetation. They dig burrows to which they retreat when threatened. The diet consists of insects (especially beetles, roaches, and ants) and spiders. Females lay eggs in soil in the summer that hatch in late summer or fall.

CONSERVATION ISSUES The species does not appear to compete with native lizards but could be problematic in areas of overlap with the six-lined racerunner.

How do you identify a giant whiptail?

BODY PATTERN AND COLOR
brown back and gray sides both have light-colored spots; blue belly; reddish tail

DISTINCTIVE CHARACTERS
large size and yellowish to white spots

DISTRIBUTION IN THE SOUTHEAST
Miami-Dade County, Florida

SIZE

15"

Giant ameivas reach lengths of more than two feet and have several established populations in Miami-Dade County, Florida.

How do you identify a giant ameiva?

BODY PATTERN AND COLOR
brown or greenish; belly blue in males, white in females

DISTINCTIVE CHARACTERS
streamlined appearance; fast runners; may run on hind legs; males develop large jowls

DISTRIBUTION IN THE SOUTHEAST
Miami-Dade County, Florida

SIZE

2'

Giant Ameiva

Ameiva ameiva

FAMILY Teiidae

DESCRIPTION These large lizards with heavy jowls reach a maximum total length of more than 2 feet. The giant ameivas found in Florida are predominantly greenish and brown. Males have a blue belly; females have a white belly.

VARIATION AND TAXONOMIC ISSUES Lizard biologists believe that at least two subspecies have been introduced into Florida and presumably intergrade when their populations come in contact. Giant ameivas are also called jungle runners.

ORIGIN AND DISTRIBUTION The species comes from Panama and South America. Several populations are known in Miami-Dade County, Florida; some have been present since the 1950s.

NATURAL HISTORY Like their smaller relatives, the six-lined racerunners, giant ameivas are ground dwellers that are active only during the day in open areas. They eat a variety of insects and snails and will presumably eat spiders and small vertebrates.

CONSERVATION ISSUES The species might compete with the six-lined racerunner in areas where their ranges overlap.

Argentine giant tegus are large terrestrial lizards that dig underground burrows.

How do you identify an Argentine giant tegu?

Argentine Giant Tegu

Tupinambis merianae

FAMILY Teiidae

DESCRIPTION As their common name implies, these lizards can be very large; many exceed 4 feet in total length. The generally black body has light speckles and spots that may form bands on the body and tail. The young are green with dark markings for several weeks after hatching. Small, raised scales on the back and sides give them a pebbled appearance.

VARIATION AND TAXONOMIC ISSUES Argentine black-and-white tegu is another common name.

ORIGIN AND DISTRIBUTION The species originated in South America. The U.S. Geological Survey has records of Argentine giant tegus from Okeechobee, Polk, and Hillsborough counties in Florida.

NATURAL HISTORY Tegus are terrestrial and dig large underground burrows to which they retreat for safety and during cold weather. In Florida, they have been reported from scrub oak habitat. They are omnivorous, eating fruit as well as insects.

CONSERVATION ISSUES The species is not known to compete with any native lizards, but a large burrowing carnivore would probably have an effect on other species if it became widely distributed. Giant tegus are reported to use the burrows of gopher tortoises.

BODY PATTERN AND COLOR head, body, limbs, and tail black with bold whitish markings; juveniles greenish with dark markings

DISTINCTIVE CHARACTERS scales on body are raised, giving pebbly appearance

DISTRIBUTION IN THE SOUTHEAST Okeechobee, Polk, and Hillsborough counties, Florida

SIZE

4'

The brown mabuya is typically a ground dweller found in areas of heavy ground vegetation.

How do you identify a brown mabuya?

BODY PATTERN AND COLOR
body tan or olive with yellowish stripe on sides; tail yellowish

DISTINCTIVE CHARACTERS
smooth scales; live-bearer

DISTRIBUTION IN THE SOUTHEAST
Miami-Dade and Lee counties, Florida

SIZE

8"

Brown Mabuya   *Mabuya multifasciata*

FAMILY Scincidae

DESCRIPTION These medium-sized lizards (up to 7 or 8 inches in total length) have a tan to olive body with a yellowish stripe along each side and a yellowish tail.

VARIATION AND TAXONOMIC ISSUES Other common names include many-lined grass skink, bronze skink, sun skink, and golden skink.

ORIGIN AND DISTRIBUTION The brown mabuya is from India, New Guinea, and Southeast Asia. A population is present in Florida on the grounds of the National Tropical Botanical Garden in Coconut Grove, Miami-Dade County; the species has also been reported from Fort Myers (Lee County) on the Gulf Coast.

NATURAL HISTORY Brown mabuyas prefer heavily vegetated areas that provide both cover and open spots for basking. They eat mostly insects but also occasionally eat fruit. Larger individuals may eat small vertebrates. Springtime mating has been observed in Florida. Females bear live young rather than laying eggs; this species is the only livebearing lizard established in the Southeast.

CONSERVATION ISSUES Its restricted geographic distribution limits the brown mabuya's potential impact on native fauna, although individuals might compete with or eat native skinks in areas of overlap.

Tokay geckos, which have become established in many locations in Florida, reach lengths of a foot or more.

Tokay Gecko

Gekko gecko

FAMILY Gekkonidae

DESCRIPTION Tokay geckos are the most common and widespread large geckos in the world. Adults are often more than a foot long and have thick, muscular bodies and big jaws. Males are larger than females and are more brightly colored. The basic body color is light to dark gray with a bluish tinge. The back, sides, tail, and head have a beaded texture and are covered with orange and red spots and blotches. The yellowish brown eyes have distinct vertical elliptical pupils.

VARIATION AND TAXONOMIC ISSUES The genus *Gekko* includes more than 20 species, but no subspecies of *Gekko gecko* are generally recognized. Rather than deriving from a person or place-name, the name "tokay" imitates the loud vocalizations made by this species.

ORIGIN AND DISTRIBUTION The tokay gecko, which originated in India and Southeast Asia, is widely distributed in Florida. Well-established colonies occur in urban areas from Miami and the Florida Keys north to Tampa–St. Petersburg, and individuals or colonies have been observed farther north in Gainesville and Tallahassee. The eggs are adhesive and able to tolerate dry conditions, and eggs attached to lumber or other products imported

How do you identify a tokay gecko?

BODY PATTERN AND COLOR body bluish gray with orange or red blotches

DISTINCTIVE CHARACTERS large, yellowish eyes with vertical pupils

DISTRIBUTION IN THE SOUTHEAST urban areas throughout Florida

SIZE

12"

from India and Southeast Asia may be the source of some populations. Disenchanted pet owners who purchased a "cool" pet that turned into a savage, biting adversary may be another source of introduction. Homeowners have been known to release individuals in the hope that these large geckos will provide insect pest control. The tokay gecko is also established in Hawaii.

NATURAL HISTORY Introduced tokay geckos are associated with urban areas, characteristically with buildings, other man-made structures, or large trees that have crevices in which they can hide during the day. They are strictly nocturnal and eat roaches and other insects associated with human habitations. They also eat other lizards, mice, small birds, and even small snakes (one ate a corn snake that was longer than the gecko [about 10 inches]). Male tokay geckos vocalize at night, possibly to attract mates, to deter other males from entering a male's territory, or both. Unlike most smaller lizards, tokays can deliver a severe bite with their viselike jaws and needlelike teeth.

CONSERVATION ISSUES The impact of this species on native lizards or other fauna will probably be minimal because of its association with urban rather than natural habitats.

Tokay geckos on Plantation Key, Key Largo, Florida, hide in the crevices of a tree during the day.

The body color of the female yellow-headed gecko is a mottling of brown and gray.

The largest yellow-headed geckos are less than 4 inches long. Adult males may have a reddish-brown head.

Yellow-headed Gecko *Gonatodes albogularis*

FAMILY Gekkonidae

DESCRIPTION Yellow-headed geckos are minuscule, dark lizards. Females are a mottled mix of brown and grayish brown, and adult males are bluish gray with a contrasting rust-colored head and neck. The largest ones are 3.5 inches long.

VARIATION AND TAXONOMIC ISSUES Two subspecies (*G. a. fuscus* and *G. a. notatus*) are reported to have been introduced into Florida.

ORIGIN AND DISTRIBUTION This tropical American species is known in the Southeast from the Lower Keys and Miami, Florida, where small, isolated populations of a few individuals occur. The well-known herpetologist Dr. Archie Carr found the first U.S. specimens in Key West, Florida, in 1939, but yellow-headed geckos have never occurred in the state in large numbers. For at least a decade after the mid-1990s, no individuals were found in Florida. The reason for the population decline is unknown. The species was still known from isolated locations in southern Florida in the early 2000s.

NATURAL HISTORY Yellow-headed geckos are most active during the day and are most commonly associated with man-made structures.

CONSERVATION ISSUES Yellow-headed geckos do not appear to compete with native lizards.

The common house gecko is native to Southeast Asia but has become established in coastal regions through much of the world. This individual is from Costa Rica.

How do you identify a common house gecko?

BODY PATTERN AND COLOR
body grayish brown; belly white

DISTINCTIVE CHARACTERS
glandular pores visible on inside of thighs

DISTRIBUTION IN THE SOUTHEAST
Monroe, Lee, and Miami-Dade counties, Florida

SIZE

4"

Common House Gecko   *Hemidactylus frenatus*

FAMILY Gekkonidae

DESCRIPTION House geckos are small, grayish brown lizards with a white belly. The presence of visible femoral pores (glands on the underside of the hind legs) is distinctive. The Mediterranean gecko, which can be similar in appearance, lacks femoral pores.

VARIATION AND TAXONOMIC ISSUES The house gecko is also called the chitchat because of its nighttime vocalizations, and the tropical house gecko because of its widespread introductions in many tropical cities.

ORIGIN AND DISTRIBUTION First reported as abundant around buildings in Key West, Florida, in the 1990s, this Southeast Asian species was later discovered in Fort Myers (Lee County) and Homestead (Monroe County). Releases from pet stores are believed to be the source of the house geckos introduced in all three areas. Determination of whether this species will be able to persist in competition with other introduced lizards will require several years. Introduced populations have also become established in Texas and Hawaii.

NATURAL HISTORY Like many of the other introduced geckos, the house gecko is associated with man-made structures, including buildings of wood or stone. At the first hint of a threat, they race up a wall and dart into a crevice. They eat a variety of small insects that they catch at night.

CONSERVATION ISSUES House geckos are not known to compete with native lizards or to have an impact on any other native fauna.

Common house geckos are found in and around buildings in residential or urban areas.

Common house geckos are small with grayish brown coloration.

The introduced Indo-Pacific gecko has become established in several locations throughout Florida.

How do you identify an Indo-Pacific gecko?

BODY PATTERN AND COLOR
body color pale yellow to brown or black with tiny white dots; belly cream to yellow; undersurface of tail orange

DISTINCTIVE CHARACTERS
skin on body smooth without tubercles; body translucent

DISTRIBUTION IN THE SOUTHEAST
Widespread in urban areas of Florida from the Lower Keys north through most of the lower half of the peninsula and into the Panhandle

SIZE

5"

Indo-Pacific Gecko   *Hemidactylus garnotii*

FAMILY Gekkonidae

DESCRIPTION A single Indo-Pacific gecko can vary in coloration from pale yellowish to dark brown or black with a sprinkling of small white dots on the back, legs, and tail. The belly is yellow, and the undersurface of the tail has an orange tinge. The maximum total length is about 5 inches. Indo-Pacific geckos make a chirping sound when another gecko of the same or another species moves into their vicinity. They lack the tubercles present on Mediterranean geckos, and the coloring of the belly and tail are distinct from other introduced geckos in Florida.

The belly of this Indo-Pacific gecko is yellow.

162 • *Indo-Pacific Gecko*

VARIATION AND TAXONOMIC ISSUES The species is parthenogenetic (that is, females reproduce without being fertilized by a male), producing only female offspring that are genetic replicas of the mother. No males have ever been discovered.

ORIGIN AND DISTRIBUTION The Indo-Pacific gecko, which originated in Southeast Asia and the East Indies, has been one of the most successful of the exotic lizard species in Florida since the early 1960s. Established colonies lasting more than a decade have been documented in at least 15 counties in the Florida peninsula, and 15 additional counties have had colonies for at least a few years. The species is also established in parts of Hawaii, Texas, and possibly in Alabama.

NATURAL HISTORY The parthenogenetic reproductive system allows a single Indo-Pacific gecko to found a population. These geckos are apparently capable of hitchhiking on nursery plants and lumber, and are therefore as-sured of opportunities for coloniza-tion throughout the Southeast and other areas of the United States. The species thrives around build-ings and other man-made struc-tures but can successfully colonize some natural habitats. The diet consists predominantly of insects. Although active primarily at night around lights where they can catch nocturnal insects, Indo-Pacific geckos have also been observed out and active during the day, especially on cloudy days.

The Indo-Pacific gecko reproduces parthenogeneti-cally. Females produce exact genetic replicas without fertilization by a male.

CONSERVATION ISSUES Indo-Pacific geckos do not appear to compete with native lizards.

Wood slaves usually display small warts along the body.

How do you identify a wood slave?

BODY PATTERN AND COLOR
light to dark brown with darker bands on back and tail

DISTINCTIVE CHARACTERS
small warts on body

DISTRIBUTION IN THE SOUTHEAST
numerous localities in the Keys and the southern half of the Florida peninsula

SIZE

5"

Wood Slave

Hemidactylus mabouia

FAMILY Gekkonidae

DESCRIPTION These medium-sized (about 5 inches total length), brownish lizards can be light (almost white) or dark, with darker chevrons on the back and crossbands on the tail. Small warts are usually visible on the body.

VARIATION AND TAXONOMIC ISSUES Common names include tropical gecko, African house gecko, Ameriafrican house gecko, and tropical house gecko.

ORIGIN AND DISTRIBUTION The wood slave originated in central and southern Africa but is now widely distributed as an introduced species in the Caribbean and in tropical South America. It has been found on most of the Florida Keys and at scattered locations

The wood slave is an introduced gecko that has established colonies on most of the Florida Keys and in many urban areas in southern Florida.

Populations of the wood slave can be found in various locations throughout southern Florida, including the Keys.

in the southern and central Florida peninsula since the early 1990s. Many populations are well established, and the species has displaced other introduced geckos—including the Mediterranean, house, Indo-Pacific, and ashy geckos—in some areas. Among the possible explanations for the comparative increase in wood slave populations is their greater success in direct or indirect competition for resources, higher rate of predation on other lizards, and more aggressive behavior when encountering lizards of other species.

Like other introduced geckos, the wood slave is active at night and eats mostly insects.

NATURAL HISTORY Wood slaves thrive in and around buildings in urban areas but are also found in trees and in association with pine woods. They are active primarily at night from May to October. Their reported diet items are mostly insects—flies, earwigs, moths, and caterpillars—and spiders. Females typically lay several clutches of two eggs during the year. As noted, they can decrease the size of or eliminate populations of other introduced gecko species. Ironically, reported predators are also introduced, namely Cuban treefrogs and tokay geckos.

CONSERVATION ISSUES Wood slaves probably have minimal competitive or predatory impact on native lizards but do interact with several exotic species in Florida.

Small populations of the Asian flat-tailed house gecko have been reported from several Florida counties, and all have been associated with buildings in the vicinity of pet dealerships.

How do you identify an Asian flat-tailed house gecko?

BODY PATTERN AND COLOR
body and tail brown with darker brown markings; belly white

DISTINCTIVE CHARACTERS
webbed toes; flat body and tail; flap of loose skin along sides

DISTRIBUTION IN THE SOUTHEAST
Alachua, Pinellas, Hillsborough, Lee, and Miami-Dade counties, Florida

SIZE

5"

Asian Flat-tailed House Gecko

Hemidactylus platyurus

FAMILY Gekkonidae

DESCRIPTION These small to medium-sized (4–5 inches in total length) geckos have a brownish body with darker brown, often intricate, markings on the body and tail and white undersides. Their webbed toes, the flap of loose skin that forms a fold along the sides, and the relatively flattened body and tail distinguish them from other introduced geckos in Florida.

VARIATION AND TAXONOMIC ISSUES The specific point of origin of introduced Asian flat-tailed house geckos is unknown, and the particular subspecies present may vary from one locality to the next. Another common name is flat-tail gecko. Some authorities place this species in the genus *Cosymbotus*.

ORIGIN AND DISTRIBUTION The species' natural range covers a vast area of Asia and the Indo-Pacific, including southern China, New Guinea, the Philippines, and India. Small colonies have been reported from widely spaced sites in Florida that include Alachua (Gainesville), Pinellas (Clearwater), Hillsborough (Tampa), Lee (Fort Myers), and Miami-Dade (Homestead) counties. All known colonies have been associated with buildings where pet dealerships had operated, an indication that the lizards had escaped or had been released intentionally.

NATURAL HISTORY Little is known about the ecology in Florida. Asian flat-tailed house geckos are most active at night and eat insects but seldom venture away from the sides of buildings. Ringed wall geckos are known predators and may have eliminated the Asian flat-tailed house geckos in Fort Myers. Individuals make a clicking sound.

CONSERVATION ISSUES Asian flat-tailed house geckos are localized and restricted to urban areas and are unlikely to have an appreciable effect on native lizards.

Did you know?

Both sexes of some lizards vocalize. Male geckos make chirping, barking, or twittering sounds to advertise their presence to females and declare their territory to other males.

The Asian flat-tailed house gecko has established colonies in several Florida counties, but the specific country of origin for these introduced populations is unknown.

Mediterranean geckos occupy many southeastern cities. This one is in New Orleans.

How do you identify a Mediterranean gecko?

BODY PATTERN AND COLOR
whitish, pink, or tan, with faint darker markings on the body

DISTINCTIVE CHARACTERS
tubercles on body and limbs; belly transparent, with eggs visible inside females

DISTRIBUTION IN THE SOUTHEAST
widespread in urban areas throughout Florida and also in cities in Georgia, South Carolina, Virginia, Alabama, Mississippi, and Louisiana

SIZE

5"

Mediterranean Gecko   *Hemidactylus turcicus*

FAMILY Gekkonidae

DESCRIPTION Mediterranean geckos are small to medium-sized (4–5 inches in total length) lizards. The semi-translucent body is pinkish, tan, or white with faded dark markings. The body is so transparent below that eggs are visible inside females. Distinct tubercles are present on the body, legs, and tail. Like many other geckos, Mediterranean geckos lack eyelids.

VARIATION AND TAXONOMIC ISSUES Of the recognized subspecies, the one introduced into the United States is considered to be *H. t. turcicus,* sometimes called the Turkish gecko.

Mediterranean geckos lack eyelids.

ORIGIN AND DISTRIBUTION The Mediterranean gecko is native to an area ranging from southern Europe westward to Turkey and India, including most of the Mediterranean and Red Sea regions. Known from Florida since about 1910, the species has probably been introduced into every other southeastern state as well and has established populations in one or more cities in most of them, including Georgia, South Carolina, Virginia, Alabama, Mississippi, and Louisiana. The species also occurs in the southern parts of all the southwestern states, Arkansas, and Maryland, generally in cities where houses, apartment buildings, or warehouses provide protection from cold temperatures.

NATURAL HISTORY Mediterranean geckos are active almost exclusively on warm nights from early evening to sunrise, often around lights likely to attract insects. They hide during the day in small crevices in concrete or behind wooden walls and light fixtures. Small insects are the primary

Small tubercles are obvious on the body of a Mediterranean gecko.

prey. In areas where their ranges overlap, Cuban treefrogs are probably predators. Mediterranean geckos mate from March to July, and females lay eggs (one or two at a time) from April to August in Florida. The mating and egg-laying seasons are more truncated to the north. Males make a peeping sound that warns off other males or signals females during the breeding season.

CONSERVATION ISSUES Mediterranean geckos are restricted to urban and suburban settings and are not likely to compete with any native lizards in the Southeast.

A small introduced population of Bibron's gecko has been reported from Bradenton, Florida, but the species is not known from other localities.

How do you identify a Bibron's gecko?

BODY PATTERN AND COLOR
body brownish gray with black and white bands and other markings

DISTINCTIVE CHARACTERS
distinctive tubercles on body

DISTRIBUTION IN THE SOUTHEAST
Manatee County, Florida

SIZE

6"

Bibron's Gecko   *Chondrodactylus bibronii*

FAMILY Gekkonidae

DESCRIPTION Bibron's gecko is a medium-sized (about 5–6 inches total length) gecko. The brownish gray body has black and white markings and is covered with tubercles. A dark line runs from the front of the mouth through the eye to the back of the head.

VARIATION AND TAXONOMIC ISSUES This species, formerly placed in the genus *Pachydactylus,* is also called Bibron's thick-toed gecko.

ORIGIN AND DISTRIBUTION A colony of this South African species that some authorities believe was intentionally introduced has thrived in Bradenton (Manatee County), Florida, since the 1970s, but the species is not known to have established populations elsewhere.

NATURAL HISTORY Introduced Bibron's geckos are associated with buildings and man-made structures and are typically active at night. They are insect eaters.

CONSERVATION ISSUES The species is known only from an urban setting and would not compete with native lizards.

Many lizard biologists consider the Madagascar day gecko to be one of the most beautiful lizards in the world.

Madagascar Day Gecko *Phelsuma madagascariensis*

FAMILY Gekkonidae

DESCRIPTION These large (often more than 10 inches in total length), vividly colored green lizards have red spotting on the back and a bold, distinctive red stripe on the face in front of each eye. The underside is white or yellowish. Many people consider the Madagascar day gecko to be one of the most beautiful lizards in the world.

VARIATION AND TAXONOMIC ISSUES The subspecies believed to have been introduced in Florida is *Phelsuma m. madagascariensis*. Another common name is giant day gecko.

Juvenile Madagascar day geckos are green like the adults.

How do you identify a Madagascar day gecko?

BODY PATTERN AND COLOR body vibrant green with red spots above; belly whitish; red line on face

DISTINCTIVE CHARACTERS body has pebbly appearance and texture; outer layer of skin peels off immediately if animal is not handled carefully

DISTRIBUTION IN THE SOUTHEAST Miami-Dade, Lee, and Broward counties, Florida

SIZE

10"

Madagascar day geckos climb trees and walls. They are intolerant of cold temperatures and may be unable to establish large populations in some areas of southern Florida where they have been reported to occur.

ORIGIN AND DISTRIBUTION No large established colonies are known, but individuals have been sighted in a few localities in Lee, Broward, Miami-Dade, and Monroe counties in southernmost Florida, including the Florida Keys. These geckos, which originated in Madagascar, are susceptible to cold temperatures and may be able to persist only in the warmest areas of the Southeast. Populations are also established in Hawaii.

NATURAL HISTORY Madagascar day geckos are diurnal, typically staying in trees or on walls where they hunt their insect prey. Their specialized toes allow them to run up the side of any surface and even across ceilings. In an unusual defense mechanism, the outer layer of skin peels off when the animal is grabbed by a predator or handled roughly. The skin regenerates, but often with some scarring. Although they are primarily insectivores, these geckos will also eat ripe fruit and nectar.

CONSERVATION ISSUES The species is not common enough to pose a conservation threat.

How do you identify an ocellated gecko?

Ocellated Gecko

Sphaerodactylus argus

FAMILY Gekkonidae

DESCRIPTION These tiny brown lizards reach a maximum length of slightly more than 2.5 inches. Although both species have a pointed snout, ocellated geckos can usually be distinguished from Florida reef geckos by the presence of numerous white spots that often form white lines on the head and shoulders. The tail is sometimes lighter in color than the body.

VARIATION AND TAXONOMIC ISSUES The introduced subspecies is considered to be *Sphaerodactylus a. argus*.

ORIGIN AND DISTRIBUTION The ocellated gecko, which originated in Cuba and Jamaica, is known in Florida only from the Lower Keys and is rare there.

NATURAL HISTORY Ocellated geckos are usually active at night, often around buildings; during the day they hide under rocks, boards, or other debris. The female lays a single egg during each nesting bout.

CONSERVATION ISSUES Ocellated geckos appear to pose no threat to other lizards.

A hatchling ocellated gecko (measured in centimeters) is shown alongside an egg.

BODY PATTERN AND COLOR brown with tiny white spots

DISTINCTIVE CHARACTERS snout distinctly pointed; white spots on head and shoulders may form lines

DISTRIBUTION IN THE SOUTHEAST Lower Keys in Monroe County, Florida

SIZE

2.5"

The ashy gecko was presumably introduced into the Florida Keys from Cuba by humans rather than by natural means.

How do you identify an ashy gecko?

BODY PATTERN AND COLOR brownish, gray, reddish, or gold body with faint reticulated pattern; juveniles with black bands on tan body and red tail

DISTINCTIVE CHARACTERS sharply pointed snout; no white lines evident on head

DISTRIBUTION IN THE SOUTHEAST Lower Keys in Monroe County, Florida

SIZE

4"

Ashy Gecko   *Sphaerodactylus elegans*

FAMILY Gekkonidae

DESCRIPTION This small (less than 4 inches in total length) lizard has a noticeably pointed snout. The highly variable basic body coloration ranges from gray to brown, reddish, or gold, with lighter spotting. Juveniles differ from adults in having dark bands across the body and a red tail.

VARIATION AND TAXONOMIC ISSUES Two subspecies are known; the one

present in Florida is *S. e. elegans*. The species is sometimes called the Cuban ashy gecko or Mac-Cleay's ashy gecko after the person who first described it.

Ashy geckos are active in early evening and at night.

ORIGIN AND DISTRIBUTION The ashy gecko is known in Florida only from the Lower Keys, where it has been a resident since the early 1900s. Whether the species arrived naturally from Cuba and should thus be considered a native species or was introduced is unresolved. The ashy gecko has not increased its range onto the Florida peninsula and is reported to have declined in some areas where it was once common.

NATURAL HISTORY Ashy geckos are active primarily in early evening and at night on man-made structures and trees such as introduced Australian pines. During the day they hide beneath bark, boards, or other debris. They eat very small insects, including ants and flies. Females lay only one egg during each nesting bout.

CONSERVATION ISSUES Ashy geckos probably have minimal or no effect on any native species.

Baby ashy geckos, with dark bands encircling the body and a red tail, look dramatically different from the adults.

Most reports of introduced ringed wall geckos in several Florida counties have been from buildings associated with pet dealerships.

How do you identify a ringed wall gecko?

BODY PATTERN AND COLOR
brown with dark bands on body and tail

DISTINCTIVE CHARACTERS
pair of white spots on each shoulder

DISTRIBUTION IN THE SOUTHEAST
Miami-Dade, Lee, and Leon counties, Florida

SIZE

6"

Ringed Wall Gecko

Tarentola annularis

FAMILY Gekkonidae

DESCRIPTION The ringed wall gecko is a relatively large (up to 6 inches in total length), thick-bodied, brownish lizard with darker bands across the body and tail. Two white spots, one behind the other, are visible on each shoulder.

VARIATION AND TAXONOMIC ISSUES Other common names include twin-spotted gecko and white-spotted gecko.

ORIGIN AND DISTRIBUTION The ringed wall gecko, which is native to Africa, is known only from Miami-Dade, Lee, and Leon counties, Florida, at widely scattered sites associated with pet dealerships. The geckos are presumed to have escaped from these facilities. No large established colonies are known.

NATURAL HISTORY Ringed wall geckos are most active at night, and in Florida they have been found on warehouse walls and associated man-made structures. They eat insects primarily but will also eat other geckos.

CONSERVATION ISSUES Ringed wall geckos could be significant predators on other lizards if populations were to reach large numbers, although native lizards are less likely to be affected than introduced ones that occupy urban areas.

A reddish head is characteristic of an adult African rainbow lizard in breeding coloration.

African Rainbow Lizard

Agama agama

FAMILY Agamidae

DESCRIPTION These rough-scaled, robust lizards reach a total length of 8–9 inches. Adult males vary in color from blue to black and have a reddish to brown head; females are brown.

VARIATION AND TAXONOMIC ISSUES The subspecies introduced in Florida is *A. a. africana*. Common agama, red-headed agama, and red-headed rock agama are other common names.

ORIGIN AND DISTRIBUTION A few established colonies of this East African native occur in southern and central Florida (Miami-Dade, Broward, Lee, Martin, Seminole, and Charlotte counties). African rainbow lizards were present in southern Florida by the early 1980s. Like many other introduced exotic species, they are not widespread or abundant but instead occur in small, persistent populations.

NATURAL HISTORY African rainbow lizards are active during the day and prefer areas with buildings and other man-made structures. Females in Florida lay as many as three clutches of seven or eight eggs each, beginning in April.

CONSERVATION ISSUES The species is known only from developed areas and does not appear to compete with native lizards.

How do you identify an African rainbow lizard?

BODY PATTERN AND COLOR
males dark gray or blue with orangish head; females brown

DISTINCTIVE CHARACTERS
stout body; large head

DISTRIBUTION IN THE SOUTHEAST
Miami-Dade, Broward, Lee, Martin, Seminole, and Charlotte counties, Florida

SIZE

9"

A blue-green color-
ation on the front of
the body is character-
istic of a male Indo-
chinese bloodsucker.

*How do you
identify an
Indochinese
bloodsucker?*

**BODY PATTERN
AND COLOR**
gray to brown; males
with blue-green head
region; females with
light stripes or bands

**DISTINCTIVE
CHARACTERS**
spiny crest on head
and down back

**DISTRIBUTION
IN THE
SOUTHEAST**
Glades and Okeecho-
bee counties, Florida

SIZE

12"

Indochinese Bloodsucker   *Calotes mystaceus*

FAMILY Agamidae

DESCRIPTION Indochinese bloodsuckers are large (almost a foot in total
length) brownish or grayish lizards. Males have blue-green coloration on
the head and shoulders, and females have stripes and bands on the body.
A crest on top of the head that runs onto the back is a distinctive feature.

VARIATION AND TAXONOMIC ISSUES The species is sometimes called the
tree agama.

ORIGIN AND DISTRIBUTION The Indochinese bloodsucker, whose natural
range is in Southeast Asia from China to India, has been reported from
Glades and Okeechobee counties in south-central Florida.

NATURAL HISTORY These big lizards typically live in trees, and at least
one colony is established in a citrus grove near Lake Okeechobee. They
eat insects.

CONSERVATION ISSUES Indochinese bloodsuckers are not known to compete with any native lizards.

The variable bloodsucker can reach lengths of more than 18 inches.

How do you identify a variable bloodsucker?

Variable Bloodsucker

Calotes versicolor

FAMILY Agamidae

DESCRIPTION These medium-sized to large, thick-bodied, brownish lizards can reach a total length of almost 1.5 feet, with the tail being twice as long as the body. A jagged crest runs down the center of the back from head to tail. Males develop a reddish coloration on the throat, head, and shoulders during the breeding season.

VARIATION AND TAXONOMIC ISSUES Of the two described subspecies, the introduced form is *C. v. versicolor*. Oriental garden lizard is another common name for this species.

ORIGIN AND DISTRIBUTION A population of these Asian lizards has been established in St. Lucie County, Florida, since at least 1978.

NATURAL HISTORY The Florida population lives in disturbed and agricultural areas along a canal bordered by citrus groves, Brazilian peppers, and other trees where they spend most of their time. The lizards sleep in trees and other vegetation at night. They eat insects but will also consume small vertebrates, including other lizards. Each female may lay a clutch of a dozen or more eggs.

CONSERVATION ISSUES The variable bloodsucker is not known to compete with or prey on any native lizards within its limited geographic range in Florida.

BODY PATTERN AND COLOR plain brownish; head region of males red during spring breeding season

DISTINCTIVE CHARACTERS disproportionately long tail; jagged crest down center of back and especially conspicuous behind head

DISTRIBUTION IN THE SOUTHEAST St. Lucie County, Florida

SIZE

18"

Butterfly lizards, which have been introduced into a residential area in southern Florida, can reach lengths of more than 18 inches.

How do you identify a butterfly lizard?

BODY PATTERN AND COLOR greenish or yellowish brown body with orange and black bars on the sides

DISTINCTIVE CHARACTERS dig burrows and run to them when threatened

DISTRIBUTION IN THE SOUTHEAST Miami-Dade County, Florida

SIZE

18"

Butterfly Lizard

Leiolepis belliana

FAMILY Agamidae

DESCRIPTION Butterfly lizards are medium to large (up to 1.5 feet total length), greenish or yellowish brown lizards with yellow spots. The lower sides have alternating orange and black bars.

VARIATION AND TAXONOMIC ISSUES Of the described subspecies, the introduced form in Florida is believed to be *L. b. belliana*.

ORIGIN AND DISTRIBUTION The butterfly lizard is native to Myanmar and Thailand. An established population has been known from a residential area in Miami-Dade County, Florida, since 1992.

NATURAL HISTORY Butterfly lizards are terrestrial and active during the day when temperatures are high. They dig underground burrows that can be more than a foot deep where they retreat during cold weather, when threatened, and at night. The diet includes a variety of insects, small crabs, and vegetation.

CONSERVATION ISSUES Butterfly lizards are not known to compete with any native lizards and are unlikely to be much of a problem in a residential area.

Veiled Chameleon

Chamaeleo calyptratus

FAMILY Chamaelonidae

DESCRIPTION These large (1–2 feet total length), impressive creatures have the classical chameleon shape—an arched body that is much higher than wide—a curled prehensile tail, and eyes on independent turrets that can look in different directions simultaneously. Individuals can change color, but not in the traditional chameleon fashion of matching their background. The greenish body, which bears a few broad, brightly colored bands, is marked in shades of green, yellow, orange, black, white, and even blue. A bony casque extends several inches above the head like a large crown. Males are larger than females, more brightly colored, and have a larger casque.

VARIATION AND TAXONOMIC ISSUES The species is also known as the Yemen chameleon.

ORIGIN AND DISTRIBUTION The veiled chameleon comes from Saudi Arabia and Yemen. An established population has been reported in Lee County on the southern Gulf Coast of Florida, and there is a single record from Collier County. The species has also become established in Hawaii.

How do you identify a veiled chameleon?

BODY PATTERN AND COLOR body greenish with orange, black, and yellow markings and light-colored bands

DISTINCTIVE CHARACTERS thick, arched body shape; curled tail; crownlike structure (casque) above head; eyes operate independently

DISTRIBUTION IN THE SOUTHEAST Lee County, Florida

SIZE

2'

Female veiled chameleons can be distinguished from males by their duller coloration and smaller casques.

NATURAL HISTORY Veiled chameleons can be found on vegetation in urban settings—juveniles typically in high grass and adults in trees or vines. Insects are the primary food. Females can lay more than 30 eggs at least three times a year and have been known to lay single clutches of up to 100 eggs.

CONSERVATION ISSUES The impact of this species on native lizards or other fauna has not been determined.

Although not native to the United States, the common wall lizard is a popular animal found in two counties in northern Kentucky.

Common Wall Lizard

Podarcis muralis

FAMILY Lacertidae

DESCRIPTION This small (5.5–7 inches long) lizard has a flattened body, small scales, and a pointed snout. Coloration and markings are extremely variable across the species' natural range. Typically, the body is gray to brown and the tail has lateral striping. The back may or may not have apparent stripes. The belly is usually lighter than the back with varying amounts of red, orange, or yellow. The throat is a much lighter gray or cream with varying amounts of dark pigmentation.

VARIATION AND TAXONOMIC ISSUES The species is widespread throughout its European range but typically is not referred to with subspecific designations.

ORIGIN AND DISTRIBUTION The natural range of the common wall lizard covers a large portion of Europe, including central Spain, Belgium, and the Netherlands. In the United States, the species is established in one southeastern state (Kentucky) as well as farther north in Indiana and Ohio. These populations are probably descendants of 10 specimens that were intentionally released in the Cincinnati area by George Rau (a member of the Lazarus department store family) in the early 1950s. The original

How do you identify a common wall lizard?

BODY PATTERN AND COLOR
gray or brownish body and tail, sometimes with stripes; light-colored throat

DISTINCTIVE CHARACTERS
very small scales; pointed snout

DISTRIBUTION IN THE SOUTHEAST
Kenton and Campbell counties, Kentucky

SIZE

7"

Populations of the common wall lizard in Kentucky and Ohio are believed to be the descendants of ten specimens released in the 1950s.

specimens were collected from Lake Garda near Milan, Italy. The population in Kentucky is located in Kenton and Campbell counties.

NATURAL HISTORY The common wall lizard thrives in urban areas where rock piles and other hiding places are available. It is an adept climber that will race up tree trunks or stone walls into crevices. The diet includes various insects and other invertebrates.

CONSERVATION ISSUES Although not a native species, the common wall lizard is protected by the Ohio Department of Natural Resources. It is not protected in Kentucky and probably has little ecological impact on other lizards.

Nile monitors are ecologically versatile, able to walk overland, swim in freshwater or salt water, climb trees, and dig burrows.

How do you identify a Nile monitor?

BODY PATTERN AND COLOR body brown with yellow spots that form bands

DISTINCTIVE CHARACTERS neck long, snakelike; tongue long, forked; tail flattened

DISTRIBUTION IN THE SOUTHEAST Lee County, Florida

SIZE

5'

Nile Monitor

Varanus niloticus

FAMILY Varanidae

DESCRIPTION Nile monitors have characteristics typical of other monitors, including a long neck and tail and an obvious forked tongue. They have long claws and often stand erect on their hind feet, using the large tail for balance. Nile monitors are the largest lizards in the United States. Adults are commonly more than 5 feet long, and the maximum total length may exceed 7 feet. The long, continually active forked tongue and dark body with several yellowish bands formed by yellow or gold spots that encircle the body are characteristic of the species. The spots look like coins, and the species is known as the money monitor in some of its native range. The tail is flattened for swimming.

VARIATION AND TAXONOMIC ISSUES Two subspecies have been described: *V. n. niloticus* and *V. n. ornatus*. Some taxonomists do not consider them distinctive enough to warrant subspecific status; others consider them separate species.

ORIGIN AND DISTRIBUTION Isolated sightings of this African native have been reported in Florida since the early 1980s. Nile monitors were sighted in the 1990s and early 2000s in Miami-Dade, Broward, Orange, Alachua, De Soto, and Collier counties, but there is no definitive evidence that breed-

Of all introduced lizards, the Nile monitor probably poses the greatest threat to native Floridian wildlife.

Did you know?

The Komodo dragon of Southeast Asia can be almost 10 feet long and includes humans among its prey.

ing populations were present. A population of Nile monitors has been established in Lee County (Cape Coral) since the early 1990s and is reported to be rapidly expanding its distribution to the surrounding mainland and barrier islands.

NATURAL HISTORY A notable feature of this large predator is its adaptability to both terrestrial and aquatic habitats. In Africa, Nile monitors are essentially semiaquatic but also use a wide array of upland habitats; they can climb trees, and they dig burrows for shelter and nesting. In Florida, they have been seen in residential areas and alongside or in canals, lakes, and tidal creeks. Nearly any invertebrate or vertebrate that this carnivorous lizard can capture and subdue becomes a meal. The diet in Florida includes insects, mollusks, crabs, shrimp, fish, amphibians, lizards and other reptiles and their eggs, birds and their eggs, and mammals. Nile monitors are said to be very intelligent and to hunt cooperatively. They will also scavenge dead animals and prowl through human garbage for food and are probably a potential hazard for small pets outdoors.

Populations of Nile monitors can expand rapidly and may reach densities of more than 200 adults per square mile. They begin reproducing before they are 3 years old and can lay as many as 60 eggs in a clutch.

CONSERVATION ISSUES With its broad diet and predatory nature coupled with its large size, adaptability, rapid growth, high fecundity, and intelligence, the Nile monitor lizard probably poses a greater threat to Florida's native wildlife, including birds and mammals, than any other introduced lizard. Nile monitors are known to eat burrowing owls, a federally listed endangered species in Florida.

Other Exotic Lizards

The list of exotic lizard species that could potentially arrive in southern Florida as escapees from the pet trade, intentional releases, or accidental hitchhikers on transport ships or aircraft is extensive. Most temperate zone or tropical lizards have the potential to persist for one or more generations in southern Florida's subtropical to tropical climate if mating pairs are available. Under optimal conditions of sufficient prey, suitable refuge and egg-laying sites, and limited predation by or competition with native or introduced species, creation of an established population is possible.

The fact that several species of exotic lizards and other reptiles in southern Florida have realized rapid population growth and long-term persistence makes the likelihood of other species appearing over the next few decades high. Likewise, some of the current representatives that have survived for several decades may disappear. Because colonization patterns and success are not easy to predict, we have chosen only to mention some of these exotic species rather than provide full accounts. Some of the species listed below have been reported outside captivity but none seem to have established sustainable populations. Their long-term future in Florida and other parts of the southeastern United States is uncertain at this time, but all are represented in the pet trade and are potential colonizers.

FAMILY AGAMIDAE

Bearded Dragon
Pogona vitticeps
Origin: Australia

FAMILY SCINCIDAE

Blue-tongue skink
Tiliqua scincoides
Origin: Australia

Fire-sided skink
Mabuya perrotetii
Origin: Africa

FAMILY IGUANIDAE

Emerald swift
Sceloporus malachiticus
Origin: Central America

FAMILY VARANIDAE

Savannah monitor
Varanus exanthematicus
Origin: Africa

FAMILY GEKKONIDAE

Moorish Gecko
(Common wall gecko)
Tarentola mauritanica
Origin: Mediterranean region

Fat-tailed gecko
Eublepharis macularius
Origin: Africa

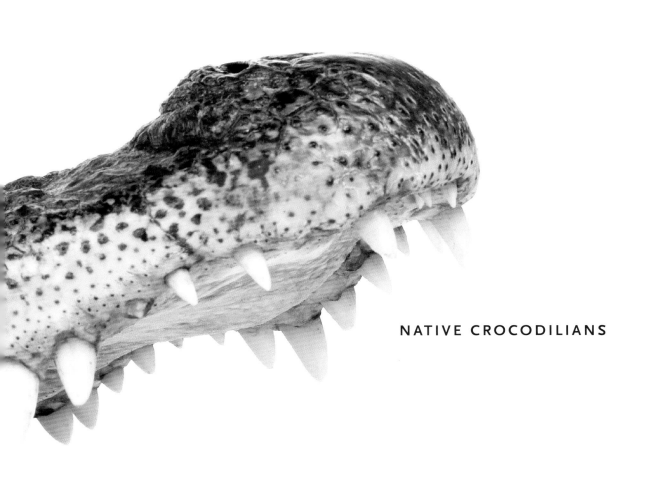

NATIVE CROCODILIANS

A male alligator (foreground) gets much longer and heavier and has larger jowls compared to the adult female.

American Alligator
Alligator mississippiensis

FAMILY Alligatoridae

DESCRIPTION American alligators have a broad, rounded snout with some teeth visible when the mouth is closed. The biggest adults are black, but some moderately large individuals retain remnants of the pale yellowish crossbands of hatchlings on the body and tail. The underside is pale white to cream colored. The eyes are yellow with elliptical black pupils. Male alligators get appreciably larger than females in both weight and length. The largest documented sizes in recent history are between 14 and 15 feet, but the all-time record in the reptilian literature is a 19-foot, 2-inch male purportedly killed in the swamps of Louisiana in 1890 by E. A. McIlhenny. Many crocodilian biologists do not consider that record valid, maintaining that the largest documented records are of alligators under 15 feet.

Hatchling American alligators

WHAT DO THE YOUNG LOOK LIKE? Hatchlings are shaped like small adults and are glossy black at hatching with bright yellow-gold crossbands. They retain the juvenile black-and-yellow pattern for a year or more. Like the adults, the young are lighter underneath.

How do you identify an American alligator?

BODY PATTERN AND COLOR
black, sometimes with faded pale bands on tail

DISTINCTIVE CHARACTERS
broad, rounded snout

SIZE

9"　　9'　　14'

● ADULT MALE
● ADULT FEMALE
● HATCHLING

VARIATION AND TAXONOMIC ISSUES The American alligator has no sub-species and looks the same throughout its geographic range.

CONFUSING SPECIES Alligators could be confused with American crocodiles or with nonnative caimans in Florida, but alligators have a much broader, more rounded snout than crocodiles (which are also lighter in color), and caimans have a curved bony ridge between their eyes that alligators lack. Nile monitor lizards, which have a dark body with yellow bands, have been mistaken for small to medium-sized alligators while swimming in freshwater habitats in southern Florida, but monitors have a distinctive long, snakelike neck.

Hatchling white alligators with blue eyes, a genetic condition known as leucism that results in reduced pigmentation, have been found in Louisiana and South Carolina. Whether such individuals would be able to survive to adulthood in the wild is unknown.

DISTRIBUTION AND HABITAT Alligators occur in the Southeast from coastal North Carolina to Louisiana and continue their range into eastern and southern Texas, the extreme southeastern corner of Oklahoma, and the lower half of Arkansas. They are present in the Coastal Plain of North Carolina, South Carolina, Georgia, Alabama, and Mississippi, and throughout Louisiana and mainland Florida. Populations occur sporadically in the northernmost Florida Keys, and a population is resident in the Lower Keys. Alligators are always associated with wetland habitats, including such water sources as rivers, swamps, bayous, slow-moving streams, man-made lakes and reservoirs, and even

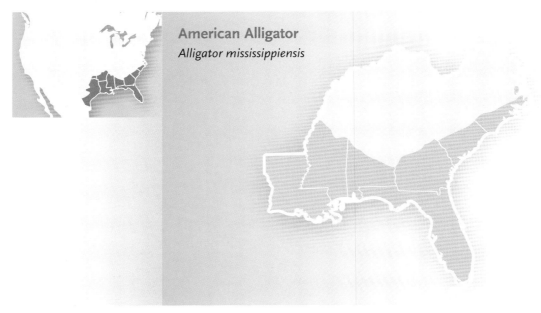

American Alligator
Alligator mississippiensis

brackish and salt water. They are found in greatest concentrations where open habitat is available that affords opportunities for basking, which they often do on logs and along the shore.

BEHAVIOR AND ACTIVITY Alligators tend to be most active in the water at night. During warm weather and on cool, sunny, windless days they bask on logs or along shorelines. They commonly feed at night but also will capture prey in the water or on the shore during the day when a meal presents itself.

Alligators in southern Florida are active during all months, but in most parts of their geographic range they become less active during the late fall and winter months when temperatures are cool. They remain underwater during cold weather but must surface occasionally to breathe. Alligators can survive unseasonably cold winter weather—including freezes—by remaining stationary for days with their nostrils above the ice and their body in the

Protected alligators reach a high population density at South Carolina's Donnelly Wildlife Management Area (top).

Alligators have bony scales (osteoderms) beneath the skin on their necks and bodies that provide protection from predators and may be used in thermoregulation (bottom).

An ancient-looking alligator covered with floating duck-weed basks alongshore.

water below, at temperatures ranging from 32 to 40°F (0–4°C). Individuals will sometimes retreat into holes in the bank during cold weather.

Alligators exhibit two distinctive behaviors during periods of drought. Some travel distances of a mile or more overland from a drying body of water to a more suitable area where standing water remains; others dig or retreat to "gator holes" in the bottom of shallow wetlands or to "caves" beneath lake, river, or stream banks. Alligator holes are often the last sources of standing water in areas of severe drought. That water serves as the last refuge for many animals, including fish, frogs, and snakes, which become a source of prey for the alligator that created the habitat.

Alligators are the most vocal reptiles in the Southeast, with a repertoire that includes grunts, hisses, and growls. Adults make deep, resonant vocalizations reminiscent of those made by bullfrogs, but a large alligator makes a rumbling sound that is much louder than that of any frog and can actually be felt by a person several feet away. Adult males do most of the bellowing, beginning in early spring, but females occasionally vocalize as well. Juveniles and young adults also communicate with low-pitched grunting sounds. When threatened, baby alligators emit high-pitched yelps to summon their mother.

FOOD AND FEEDING Alligators are carnivorous, eating whatever they can catch in the water or on shore near the water's edge, including fish, snakes, turtles, frogs, blue crabs, snails, and other invertebrates. Medium-sized to large mammals (pigs and deer), waterfowl, wading birds, and sometimes even smaller alligators are all suitable meals for adult alligators, which seize prey in their strong jaws and swallow most items whole. Animals that drink from ponds or canals—whether wild animals or pets—are in danger of falling prey to alligators. If a prey item is too large, the alligator will set its teeth in the carcass and twist its body to tear off bite-sized chunks. They also readily eat carrion and will cache leftovers and large prey items such as deer until they decompose sufficiently to be torn into edible pieces. Hatchlings and juvenile alligators feed on smaller food items such as small fish, amphibians, and invertebrates. Baby alligators will often gather around their mother when she is eating and eat scraps of food that break loose, in a behavior that has been described as incipient parental care. No other reptiles are known to feed their young.

REPRODUCTION Alligators mate in the spring. The males' characteristic bellowing can be heard at night and sometimes during the day and is thought to serve the twofold purpose of announcing their territory to other males while simultaneously attracting females. Male alligators will fight each other during the mating season, and their battles often result in severe maiming or death for the loser. Courtship is an elaborate process that includes bellowing and head slapping, and ends with copulation.

In May or June, a gravid female selects a secluded site near water and constructs a nest of vegetation about 7 feet in diameter and 2–3 feet high. She lays about 30–50 eggs (nests have been found with as many as 90 eggs) in the mound and covers them with rotting vegetation. About 65 days later, when the babies are ready to hatch, they make grunting noises while still inside the eggs. The mother alligator, who usually stays in the water near the nest and will defend it ferociously from predators, shows a tender side, responding to the cries of the entombed babies by excavating the nest, gently assisting the babies from the

Adult alligators continually lose and replace their teeth throughout their life.

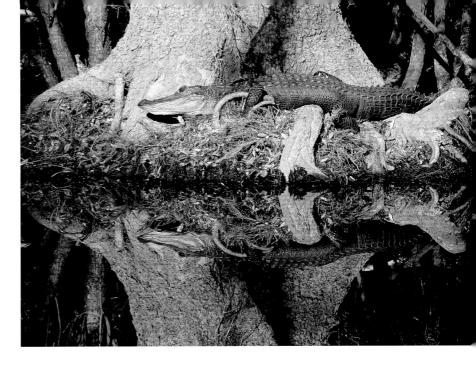

Alligators are always associated with wetland habitats.

eggs, and sometimes even carrying them to the water in her mouth. The maternal tendencies of female alligators toward their young definitely support the proposed evolutionary connection between crocodilians and birds.

The sex of baby alligators is determined by the temperature at which the eggs are incubated. Eggs kept above 93°F (34°C) become males, and those kept below 86°F (30°C) become females. Eggs incubated at intermediate temperatures may become males or females.

PREDATORS AND DEFENSE Aside from humans, predators of adult alligators are rare or nonexistent, but many animals prey on the eggs and young. Raccoons, foxes, skunks, and opossums lucky enough to find such a rich meal dig out and eat the eggs if the mother alligator is not nearby ready to make a meal out of the would-be predator. Hatchling alligators and small juveniles are eaten by large fish, raccoons, wading birds, and larger alligators, and perhaps also by the introduced Burmese pythons that are increasing in numbers in southern Florida and share much of the same habitat. Groups of hatchlings, called pods, may stay together in the same vicinity for as long as 3 years.

CONSERVATION ISSUES Alligator populations face the same threats that menace other wetland wildlife. Habitat loss due to development and alteration is probably the greatest threat now that the species is protected from wanton killing. Alligators are occasionally killed crossing roads, and unfortunately, some continue to be shot or harassed by vandals.

Interacting with Alligators

Although large adult alligators are capable of seriously injuring or even killing people, remarkably few alligator-caused human deaths occur (deer are responsible for more human deaths than alligators). Humans in the Southeast and alligators have countless interactions each year. Water skiers, boaters, and canoeists in lakes with alligators, people fishing along the shores of lakes or rivers where alligators abound, and golfers passing by water hazards where alligators live are increasingly likely to come into contact with them. People engaged in these activities are typically unaware of the hazards of being in such close proximity to a large predator or do not care. Considering the high number of opportunities for large alligators to attack humans, the actual number of deaths is quite modest. In fact, the number of alligator-caused deaths may be even lower than reported because in many situations it is unclear whether the alligator actually killed the person or scavenged an already dead body.

Anyone living around or visiting habitats where alligators occur should be aware of the following facts:

1. Feeding wild alligators is illegal and dangerous. People are far more likely to be injured if alligators become accustomed to them and no longer avoid them.

2. Alligators will come out of the water to attack a dog—even one on a leash—because they perceive it as prey. Anyone injured by an alligator while walking a dog is collateral damage, not the target of the attack.

3. Mother alligators will defend their eggs or babies in many situations, but their aggressive behavior is usually only a threat display that allows a person plenty of time to retreat from the area. Picking up baby alligators is illegal and can be dangerous because of the mother's potential protective response.

4. Swimming in areas where large alligators occur is not a safe form of recreation, especially at night.

5. Wild alligators will typically avoid adult humans, but children should be encouraged to stay a safe distance from shore when large alligators are present in the water lest they be mistaken for prey.

6. Alligators can run on land about as fast as an average person for a few yards, but they cannot sustain that speed for very long. Most people can easily outrun an alligator on land, and when being pursued are highly motivated to do so.

7. Aside from professional herpetologists and wildlife handlers, people should never try to capture wild alligators. As with snakebites, many alligator bites happen to individuals who are attempting to catch or pick up the animal.

8. State wildlife departments differ in their responses to calls about an alligator that is perceived as a nuisance or potential danger to people or pets; some states even have "nuisance 'gator" removal experts on retainer. The alligator is a protected species, however, and state game officials should be contacted to determine the proper course of action regarding alligator removal.

Baby alligators seen alongshore should not be picked up, as a protective mother may emerge open-mouthed from the water in an impressive display that could result in injury.

The American crocodile is native to southern Florida and also has an extensive geographic range in tropical America. The species does not vary noticeably in body shape throughout its range.

How do you identify an American crocodile?

BODY PATTERN AND COLOR
light or dark gray-green, sometimes with black body markings

DISTINCTIVE CHARACTERS
pointed snout with fourth lower tooth visible

SIZE

10" 12' 15'

- ● ADULT MALE
- ● ADULT FEMALE
- ● HATCHLING

American Crocodile

Crocodylus acutus

FAMILY Crocodilidae

DESCRIPTION American crocodiles have a long, narrow snout, and teeth are visible on both jaws when the mouth is closed—in particular, the fourth lower tooth is visible on each side, a trademark of all species of crocodiles. The body color is usually dark gray-green but ranges from light brownish or tan to dark olive green or brown. The back, sides, and tail, especially of younger and smaller individuals, may have black spots or bands.

WHAT DO THE YOUNG LOOK LIKE? Hatchlings resemble adults in body shape but are typically light gray or greenish in color with distinct black banding and/or spotting on the body and tail.

VARIATION AND TAXONOMIC ISSUES The American crocodile does not vary noticeably in appearance throughout its geographic range, and no subspecies are recognized.

CONFUSING SPECIES In areas of Florida where American crocodiles, alligators, and caimans could all be encountered, crocodiles can be distinguished by their much narrower snout.

DISTRIBUTION AND HABITAT The American crocodile occurs naturally in coastal areas from Vero Beach on the Atlantic side and Tampa on the Gulf side of the Florida peninsula to the lowest Florida Keys. Outside the United States, the species is found on islands of the Caribbean, including Cuba and Jamaica; in northern South America; in most of Central America; and in southern and western Mexico. American crocodiles are typically found in aquatic habitats that range in salinity from freshwater to slightly or heavily brackish or even salt water. In Florida, American crocodiles may be found in open lakes, lagoons, and river estuaries associated with Biscayne Bay, Florida Bay, and the Florida Keys and may inhabit inland mangrove swamps, canals, and ponds.

BEHAVIOR AND ACTIVITY American crocodiles are most active at night, but they also bask along shorelines during the day in areas where protected populations still exist. Because southern Florida is a subtropical habitat with few severe cold periods, crocodiles are active throughout the year, although females appear to shift their habitat during the nesting period. Adults whose

movements were monitored moved more than 10 miles along waterways between aquatic habitats, and recent hatchlings will move a mile or more from their nesting site to a suitable nursery habitat.

Hatchlings are light gray or greenish in color with distinctive black bands and/or spots.

American Crocodile
Crocodylus acutus

FOOD AND FEEDING American crocodiles eat any animals they can catch in and around the water, including fish, crabs, turtles, snakes, birds, and mammals. Smaller individuals eat a wide variety of invertebrates and small fish.

REPRODUCTION In Florida, courtship and mating begin in February and March in mangrove areas, and eggs are laid in April and May. Most of the known nesting areas are in southern Biscayne Bay, northern Key Largo, the northeastern area of Florida Bay, and Cape Sable in Everglades National Park. Females lay 15–55 eggs (average about 40) either in holes they dig or in the center of mounds they build of sand, peat, or marl. The female covers the nest cavity when she is finished laying eggs. Occasionally, more than one female will lay eggs in the same nest. The young hatch after about 3 months, at which time the female excavates and opens the nest to allow the hatchlings to escape to the water.

PREDATORS AND DEFENSE Known or suspected natural predators of American crocodiles include raccoons that prey on eggs in the nest. Large birds, fish, crabs, and raccoons will eat small juveniles, and blue crabs will drown and eat hatchlings. Small crocodiles could presumably fall prey to Burmese pythons, which have established populations in the Everglades. Adult crocodiles have no natural predators in Florida. Young crocodiles sometimes vocalize when threatened but not as commonly as young alligators. Crocodiles in Florida do not typically defend nests or young from humans, although they may defend them from natural predators.

CONSERVATION ISSUES In 1975 the American crocodile was officially listed as endangered by the U.S. Fish and Wildlife Service. The Florida population presumably increased over the next three decades, based on a quadrupling of the number of nests observed, and the species was downlisted to threatened in 2007. The status of the American crocodile in most other countries where it is found is more problematic and not very well documented. Continued threats to crocodiles in Florida include destruction of mangrove swamps and other brackish water habitats, deaths on highways, and intentional shooting or harassment.

American crocodiles in Florida inhabit freshwater lakes and canals and brackish habitats such as estuaries and tidal creeks. They also occasionally enter the ocean.

INTRODUCED CROCODILIAN

The bony ridge running across the snout in front of the eyes is a distinctive trait of the spectacled caiman.

Spectacled Caiman

Caiman crocodylus

FAMILY Alligatoridae

DESCRIPTION Caimans are moderate-sized crocodilians. Adults may reach a total length of 6 feet, but most of those found in Florida are smaller. The body and tail are brownish with dark banding and dark blotches. A curved, bony ridge (the "spectacle" from which the common name derives) runs across the snout in front of the eyes. Hatchlings are yellowish with black bands and spots.

BODY PATTERN AND COLOR
body and tail brown with dark bands and blotches

DISTINCTIVE CHARACTERS
bony ridge on snout in front of eyes

DISTRIBUTION IN THE SOUTHEAST
Miami-Dade County, Florida

SIZE

Although formidable in appearance with bony armor and large teeth, caimans are generally wary of humans and will readily retreat if given the opportunity.

A spectacled caiman (bottom) with its muted olive and brown body pattern can be readily distinguished from an alligator of similar size, which retains yellow banding.

A spectacled caiman (left) and an American alligator swim alongside each other at the St. Augustine Alligator Farm in Florida.

VARIATION AND TAXONOMIC ISSUES The species has also been called the brown caiman in Florida. Most authorities recognize four subspecies of spectacled caiman, although other subspecies have been proposed.

ORIGIN AND DISTRIBUTION Caimans, which are native to tropical America from Mexico southward, have been introduced to the wild in many places in the United States, probably as escaped and released pets, but breeding populations have persisted only in south Florida and Puerto Rico. In Florida, a single breeding population occurs at the edge of the Everglades in and around the city of Homestead (Miami-Dade County); breeding has been reported since the 1950s. Caimans are occasionally sighted elsewhere in Florida, but none of these are members of a viable reproducing colony. Individuals of other caiman species, including the black caiman, have been reported from various parts of southern Florida, but these species have not become established.

NATURAL HISTORY In Florida spectacled caimans are abundant only near Homestead in urban, suburban, and agricultural areas with vegetated canals. They are normally shy and will retreat to the bottom, into culverts, or into heavy vegetation when approached. Caimans typically eat aquatic invertebrates and live or dead fish, including introduced exotic species. They also prey on vertebrates that venture near or into the water, including birds, snakes, and frogs. Caimans build mound nests like those of alligators, and females will guard their nests.

Old adult caimans lose the banding pattern and yellowish body color of juveniles and may develop a solid dull gray or dark olive green coloration.

CONSERVATION ISSUES Caimans share their Florida habitat with alligators and occur close to areas where American crocodiles live and breed. Cannibalism is common in crocodilians, and each of these three species preys on the young of the others, but there is no evidence that caimans, alligators, and crocodiles have any lasting negative effect on one another. Caimans probably prey on a variety of native fauna, but the areas where they occur are already environmentally disrupted by urbanization, and one more predator is unlikely to have much additional impact. Although spectacled caimans have been breeding near Homestead for more than 50 years, no other breeding populations have been established in Florida.

Lizard biologists take notes on a captured broad-headed skink as part of an ecological study on habitat preferences of the species. One means of capturing some southeastern tree-climbing lizards is to use a noose on a string attached to the end of a rod-and-reel (leaning against tree).

People and Lizards and Crocodilians

STUDYING LIZARDS AND CROCODILIANS

People who study amphibians and reptiles are called herpetologists. Among those who consider themselves herpetologists are scientists, zookeepers, and wildlife managers. Some herpetologists study the ecology, physiology, behavior, or genetics of amphibians or reptiles, generally focusing on a particular group. Others, including many who work for state or federal agencies, concentrate their efforts on conservation, assessing the population levels and status of amphibians and reptiles in particular regions or habitats. Both native and exotic lizards and crocodilians are the targets of conservation initiatives in the Southeast. Numerous scientific papers are written each year on amphibians and reptiles. Some scientific journals are dedicated to publishing only herpetological research, while others, including conservation journals, give extensive coverage to herpetofauna.

Quite a few professional herpetologists developed their interest in amphibians or reptiles at an early age, often before they reached high school. Many adults have never lost the thrill of catching their first frog around a pond or rescuing their first box turtle crossing a road. The common childhood fascination with dinosaurs can readily lead to a natural interest in

Swimming in areas where large alligators occur is not an advisable form of recreation, especially at night. Feeding wild alligators is illegal and is the most common cause of alligator injuries to humans because the alligators become accustomed to people and no longer avoid them.

creatures that look like modern-day, living versions of animals first experienced only in books, movies, or museums.

Why Do Herpetologists Study Lizards and Crocodilians?

Herpetologists study lizards and crocodilians in many ways and for many reasons. Lizards make good subjects for behavioral experiments in the laboratory because most species are small and easy to keep in captivity, and they are nearly as easy to study in the field. An observer with binoculars inside a blind can record lizards' courtship displays, feeding behavior, and other interactions with their surroundings without disturbing them.

Because some lizards are short-lived relative to other reptiles, they are ideal for examining population responses to different environmental conditions. Field research on alligators and crocodiles has been successful in many areas because investigators can use radiotelemetry to examine their movement patterns. Crocodilians are particularly interesting to behaviorists because they exhibit complex social behavior and parental care that can be observed in field situations.

Many conservation biologists with an interest in herpetology have concentrated their field research efforts on lizards and crocodilians. The exotic species of lizards in the Southeast are of particular interest because they may ultimately affect native species of lizards and other animals in significant ways. Studies on the American alligator are especially pertinent because of efforts to protect the species while making it a valuable game animal in some areas. Studies that monitor and evaluate the status of particular species of lizards or crocodilians in the Southeast both offer insights into the basic biology of the species and can be of value to conservation biologists attempting to address active or potential environmental problems on a broader scale.

How Do Herpetologists Study Lizards and Crocodilians?

Herpetologists use a variety of techniques to study lizards and crocodilians. The measurements taken vary according to the questions being asked in the study, but most ecological, behavioral, and physiological techniques used in the field and laboratory are standard ones used for other animals. Herpetologists interested in genetics, for example, collect tissue samples for DNA analyses. Most ecological studies involve measurements of size (snout–vent length, tail length, and body weight). The sex of many species is often easy to determine from a distance because adult males tend

A lizard biologist shows the teeth of a Gila monster from Cochise County, Arizona. The Gila monster is the only venomous lizard in the United States and has a viselike grip when it bites.

to be more colorful. Many herpetologists use the mark-release-recapture method with both lizards and crocodilians to assess population size, determine individual growth rates, and estimate mortality patterns. Lizards can be individually marked by clipping the end of the toes according to a coding scheme or, for shorter-term studies, by applying paint or fingernail polish spots. Individual American alligators can be marked by clipping the scutes on the tail.

Collection techniques tend to focus on particular traits of the species being collected. Many species of lizards, especially anoles, can be captured effectively by hand. Putting down coverboards of wood or tin is an effective technique for passive collection of some species of southeastern lizards—especially skinks—that hide beneath flat surfaces. Among the more innovative techniques are using water pistols to squirt geckos on walls so that they lose their adhesive grip, using blowguns with corks to stun fast-moving lizards temporarily, and using dust shot from pistols to collect lizards for museum specimens.

Crocodilian capture techniques are not nearly as varied as those used with lizards. The most important consideration is the animal's size. Adult American alligators and most species of caimans and crocodiles must be constrained in a manner that does not permit the animal to bite or use its tail, which can be a dangerous weapon. Alligators can be attracted to a baited snare trap that is set and left overnight. A noose made of flexible cable or rope is usually the first step in securing a large crocodilian—either a free-swimming individual or one that has been caught in a snare trap. The animal's legs can be tied behind its back and duct tape or a large rubber band put around the snout. Placing a towel over the eyes often keeps the animal calm. A crocodile expert can usually capture individuals less than about 5 feet long by hand. Recent hatchlings are unlikely to bite if picked up, but a protective mother lurking nearby is always a threat. Alligators can be counted at night using eye shine reflected from a flashlight. Several studies have used radiotelemetry to examine alligator movement patterns and habitat selection. Some used a collar with a radiotransmitter and others placed the transmitter internally. The transmitter, which is set at a particular frequency, can be tracked with an antenna.

Many species of lizards make good pets and can be easily maintained in captivity. A pet owner should find out about the particular species and its caging and food needs before choosing a pet lizard.

KEEPING LIZARDS AND CROCODILIANS AS PETS

Introduction

Lizards' vivid colors, interesting body shapes, and swift movements often inspire people to keep them as pets in homes or offices. Some species fill this role well; others do not. Regardless of the species, all have certain basic needs that must be met if the animal is to stay healthy. The prospective owner should consider the species' behavior and caging and food needs when choosing a pet lizard. And as is true for most wild animals kept in captivity, ethical and conservation-related issues must also be considered.

Many species of crocodilians, especially American alligators and caimans, were once popular as household pets—while they remained small, at any rate—but the myriad federal and international laws governing endangered and threatened species have discouraged that practice today. Although zoos and other educational facilities (and in special circumstances individuals) can obtain permits to retain some species, keeping a pet alligator, caiman, or crocodile is not practical for most people.

Practical and Ethical Considerations for Pet Lizards

Anyone who wants to keep a wild-caught lizard as a pet should consider how the population of the species might be affected by removal of an individual and whether the specimen itself will do well in captivity. Herpetologists have varying opinions regarding taking lizards or other animals from the wild and keeping them as pets. Some believe that removing a single lizard will do no long-term harm to the local population. Others think that no wild lizard should ever be taken as a pet. Numerous species of lizards bred in captivity are available from legitimate pet dealers. Regardless of their attitude on how one should obtain the animals, however, many professional herpetologists agree that keeping lizards and other reptiles and amphibians as pets promotes understanding and respect for the animals and their environment.

A different dilemma arises if a pet owner no longer wishes to keep a pet lizard. In this case, other questions

Leopard geckos are one of the most popular lizard pets and can be acquired from many pet stores.

arise. How will the local lizard population be affected by the animal's release? Will a lizard fed in a cage for months be able to find and catch its own food? Has a captive lizard contracted a communicable disease that might affect other individuals in the wild population? Is it being returned to the same place where it was caught, or at least to a similar habitat, at an appropriate time of year? A pet lizard released outside its area of original capture might affect the natural genetic makeup of lizards in the area. Finally, some states actually have laws that prohibit the release of pet animals back into the wild. These are among the complex issues that anyone taking a lizard as a pet must consider.

Lizard Choices

The Southeast is home to 57 lizard species (18 native and 38 exotic, plus 2 species introduced from the western United States), each of which has special requirements for survival. The first step in choosing a pet lizard is to select an animal that lends itself to a captive environment. Larger, active lizards such as introduced iguanas and ameivas can be difficult to house because they require spacious cages and rugged accoutrements. They may fare best in outdoor cages or pens, especially if they require natural sunlight. Other species are quite shy and choose to remain hidden from view at all times (which limits their appeal as pets for most people). Many species never adapt to life in captivity, remaining nervous and refusing to eat regardless of the conditions. Others, however, such as some of the skinks, anoles, and fence lizards, require smaller enclosures and adapt well to a caged environment. A prospective lizard pet owner should thoroughly investigate information about the species before choosing it as a pet.

Feeding Lizards

Most lizards in the Southeast are carnivorous and eat invertebrates — including insects, spiders, and earthworms — as well as small vertebrates. In captivity, many thrive on a diet of crickets, mealworms, or earthworms (available at most bait stores and pet shops). Accurate information about food preferences of particular species is available in reptile journals, on Websites, and in husbandry books. Dietary supplements such as calcium and phosphorus are often required to keep pet lizards healthy and are available in pet supply stores.

Caging Lizards

A clean, well-ventilated environment in the form of an aquarium, cage, or plastic container is a must. Keep in mind that the lizard will have a lot of time to figure out a way to escape from its cage, so a secure, well-fitting lid is essential. Some species are more arboreal than others. For these lizards, vertical space may be more important than floor space, making a taller cage a better choice. Live or artificial plants add aesthetic appeal as well as providing hiding places. Rocks, logs, and moss add texture and provide a naturalistic look to a cage. A bowl will provide clean drinking water, but certain lizards (such as anoles) require a humid environment and must have their cages misted frequently with water droplets to prevent them from drying out.

Proper heating is essential for all lizards. The cage or enclosure should

Iguanas can be difficult to house because they require spacious cages and rugged accoutrements.

have a thermal gradient that allows the animals to choose their preferred temperature. They need spots where they can warm up by basking and spots where they can cool off. A hot spot 90–95°F (32–35°C) in the cage can be achieved with a floodlight or ceramic heater on one side of the enclosure. A corresponding cool spot of approximately 70°F (21°C) is also suggested. Certain species spend much of their time basking, and receiving valuable ultraviolet rays from the sun in the process, and may need at least some time outside. Many lizard enthusiasts choose to make outdoor enclosures or even mobile units (cages on wheels that can be pushed outside when the weather is appropriate) for their pets.

Collecting Lizards as Pets

Many reptiles in the Southeast, including two lizards (the sand skink and blue-tailed mole skink), are protected by the federal Endangered Species Act. State laws protect even more species and require a permit for their collection. Prospective pet owners should check and adhere to the laws related to capturing and keeping wildlife species in their state. Be aware that general collecting permits usually do not give permission to remove specimens in parks and wildlife refuges. A special permit is usually required to collect in protected areas, and typically such permission is granted only to scientific researchers. Although taking a single specimen is unlikely to cause long-lasting ecological problems, commercial collecting on a large scale can have major negative consequences for local populations.

Collecting a specimen from the wild carries great responsibility. Anyone who chooses to collect a lizard as a pet assumes the responsibility for its health and well-being for the rest of its life, which may be many years. Captive animals returned to the wild should be in robust health and should be released exactly where they were originally caught during weather conditions favorable for the species. Exotic species (from other parts of world) should never be released into the wild.

CONSERVATION OF LIZARDS AND CROCODILIANS

Addressing all of the conservation needs of lizards and crocodilians is beyond the scope of this book, but we hope that as readers become familiar with the fascinating array of species that inhabit the southeastern United States, they will want to be involved in conserving our native fauna. As has been noted before, people are most likely to care about and protect species when they can recognize them and have learned something about their ecology and behavior.

Did you know?

*The Grand Cayman blue iguana (*Cyclura lewisi*), which reaches a length of around 5 feet and can live more than 50 years, is one of the rarest lizards in the world. Some lizard biologists predict that it will be extinct in the wild by 2020.*

Lizards and crocodilians have not been spared from the threats facing other reptiles and other native wildlife in virtually all regions of the world. Modern lizards and crocodilians are the successful descendants of reptiles that first appeared on Earth more than 200 million years ago. Crocodilians have changed their general appearance and lifestyle very little since Jurassic times. Lizards are not quite as ancient as crocodilians, but on a geological time scale they are not far behind. Although both groups have lived through global climate changes, survived the threats of natural predators, and competed successfully for resources with mammals and birds, many of the world's reptiles had begun to decline in numbers by the beginning of the twentieth century, and have become increasingly imperiled on a global scale since then.

Six primary factors have been documented as causing global population declines of both reptiles and amphibians. The causes, which are not mutually exclusive, are (1) loss and degradation of natural habitats, (2) introduction of nonnative species, (3) environmental pollution, (4) disease and parasites, (5) unsustainable use due to hunting or removal for commercial purposes, and (6) global climate change. Some of these factors are directly responsible for the decline or disappearance of local populations of some species in the Southeast.

Southeastern lizards illustrate an ironic twist to standard conservation concerns about loss of biodiversity. At the beginning of the twenty-first century, three times as many species of lizards had established populations in the Southeast (primarily in Florida) than were present during earlier centuries. Most of the introduced exotic species responsible for this increased biodiversity are unlikely to develop populations outside Florida, and many of them are likely to disappear after prolonged cold snaps or from other causes. We know very little about how these new colonists may be affecting native species of lizards or other

Exposure to animals is an effective form of environmental education that leads to a conservation-oriented attitude. Little children do not have an innate fear of lizards or alligators, even larger ones. With some encouragement, most youngsters can be convinced to approach, touch, or hold a small alligator, but they should do so only when an adult familiar with alligators is present.

animals. Presumably, the natural resource base changes when a new species becomes established. But how the new suite of lizards in the Southeast will affect the natural environments through competition with or predation on other species is yet to be determined. Only one thing is certain: Nature is dynamic and change is inevitable.

The American alligator represents a conservation paradox of a different sort. The U.S. Fish and Wildlife Service (USFWS) officially listed the alligator as endangered in 1967 because it had been hunted almost to extinction for meat and skins. The species has made a remarkable comeback since then. This success story is often used as an example of how the U.S. Endangered Species Act can be effective in identifying and solving the problems confronted by a declining species. In 1987 the USFWS declared that the American alligator had recovered and turned its management over to state wildlife departments. Several southeastern states now allow limited, state-regulated hunting of alligators. Although the species is no longer on the federal list of endangered species, the USFWS continues to regulate commercial trade of skins and other alligator products because they are difficult to distinguish from skins of other crocodilians that are seriously endangered.

The most effective conservation effort for the long-term survival of American alligators and other wetland species is the protection of habitat.

Why are lizards and crocodilians important and why should we support conservation programs that protect the habitats where they live? Environmental ethics and aesthetics are among the most important reasons.

A child looks in fascination at a green anole in a residential area in South Carolina.

Beyond the fact that lizards and crocodilians are fascinating because of their origins, ecology, and behavior, especially when observed in their natural settings, most informed citizens feel that society is obligated to provide responsible stewardship for our native fauna. In addition, both groups play an important role in the natural ecosystem of the Southeast. Lizards are standard prey for many other species, including certain snakes and birds. Less well known is the fact that alligator eggs and young are important prey for mammals, birds, snakes, and turtles. American alligators and American crocodiles have a significant impact on a variety of larger prey species wherever they occur, and native lizards annually consume an astounding number of insects and other small invertebrates. A healthy ecosystem is expected to have a complex food web, and lizards and crocodilians are an important part of the mix in many regions of the Southeast.

Becoming educated about lizards, crocodilians, or any wildlife group of interest is the first step in becoming involved in effective approaches to conservation for the target group. We recommend that people interested in the reptiles of the Southeast not only educate themselves about the species but also help inform others of the steps necessary to ensure that our lizards and crocodilians remain a healthy, viable part of our natural heritage. Become a member of Partners in Amphibian and Reptile Conservation (PARC), a national conservation group that focuses on reptile and amphibian conservation, or join a local or regional herpetology group and learn about the species in your area and any conservation concerns. Most of the southeastern states have herpetological organizations, and some areas may have more than one. Become acquainted with local nature centers and museums that have nature programs.

The best way to contribute to conservation of lizards and crocodilians is to continue to learn about the species that inhabit your area. Go out and look for lizards. If you are within the range of alligators or American crocodiles, find out where you might go to observe them. Developing an awareness that these animals are around you will enhance your appreciation for them and motivate you to encourage others to help keep them an integral part of our natural environments in the Southeast.

What lizards and crocodilians are found in your state?

NATIVE SPECIES

COMMON NAME	VA	KY	TN	NC	SC	GA	FL	AL	MS	LA
WORMLIZARD										
Florida worm lizard						●	●			
ANOLES AND FENCE LIZARDS										
Green anole			●	●	●	●	●	●	●	●
Eastern fence lizard	●	●	●	●	●	●	●	●	●	●
Florida scrub lizard							●			
WHIPTAIL										
Six-lined racerunner	●	●	●	●	●	●	●	●	●	●
GLASS LIZARDS										
Slender glass lizard	●	●	●	●	●	●	●	●	●	●
Island glass lizard					●	●	●			
Mimic glass lizard				●	●	●	●	●	●	
Eastern glass lizard	●			●	●	●	●	●	●	●
SKINKS										
Little brown skink	●	●	●	●	●	●	●	●	●	●
Coal skink	●	●	●	●	●	●	●	●	●	●
Mole skink						●	●	●		
Common five-lined skink	●	●	●	●	●	●	●	●	●	●
Southeastern five-lined skink	●	●	●	●	●	●	●	●	●	●
Broad-headed skink	●	●	●	●	●	●	●	●	●	●
Florida sand skink							●			
Prairie skink										●
GECKOS										
Florida reef gecko							●			
ALLIGATOR										
American alligator				●	●	●	●	●	●	●
CROCODILE										
American crocodile							●			
TOTAL	9	8	9	12	13	15	19	13	12	12

INTRODUCED SPECIES

COMMON NAME	VA	KY	TN	NC	SC	GA	FL	AL	MS	LA
ANOLES AND FENCE LIZARDS										
Collared lizard										●
Texas horned lizard				●	●		●			
Hispaniolan green anole							●			
Crested anole							●			
Large-headed anole							●			
Bark anole							●			
Knight anole							●			
Jamaican giant anole							●			
Cuban green anole							●			
Brown anole					●	●	●	●		●
Green iguana							●			
Mexican spinytail iguana							●			
Black spinytail iguana							●			
Brown basilisk							●			
Northern curly-tailed lizard							●			
Red-sided curly-tailed lizard							●			
WHIPTAIL LIZARDS AND TEGUS										
Rainbow whiptail							●			
Giant whiptail							●			
Giant ameiva							●			
Argentine giant tegu							●			
SKINK										
Brown mabuya							●			
GECKOS										
Tokay gecko							●			
Yellow-headed gecko							●			
Common house gecko							●			
Indo-Pacific gecko							●			
Wood slave							●			
Mediterranean gecko	●				●	●	●	●	●	●
Asian flat-tailed house gecko							●			
Bibron's gecko							●			
Madagascar day gecko							●			
Ocellated gecko							●			
Ashy gecko							●			
Ringed wall gecko							●			
OLD WORLD LIZARDS										
African rainbow lizard							●			
Indochinese bloodsucker							●			
Variable bloodsucker							●			
Butterfly lizard							●			
TRUE CHAMELEON										
Veiled chameleon							●			
WALL LIZARD										
Common wall lizard		●								
MONITOR										
Nile monitor							●			
ALLIGATOR										
Spectacled caiman							●			
TOTAL	1	1	0	1	3	2	39	2	1	3

Glossary

Biodiversity The number, distribution, and abundance of species within a given area.

Brumation A period during which "cold-blooded" animals, or ectotherms, are inactive, usually during cold weather. *See also* Hibernation.

Cloaca (*adj.* cloacal) A single opening through which the urinary, digestive, and reproductive tracts exit the body.

Clutch A group of eggs laid together at one time by a single individual.

Cold-blooded A nontechnical term that refers to animals whose body temperature is largely determined by environmental conditions and the thermoregulatory behavior of that animal. *See also* Ectotherm.

Diurnal Active during the daytime.

Ecdysis The process of shedding the outer layer of skin.

Ecology The study of how organisms interact with their environment.

Ectotherm An animal whose body temperature is largely determined by environmental conditions and the thermoregulatory behavior of that animal. *See also* Cold-blooded.

Endangered Referring to a species or population that is considered at risk of becoming extinct.

Endemic Referring to a species found only in a particular geographic location and nowhere else.

Endotherm An animal that maintains a high body temperature primarily with heat generated by its own metabolism. *See also* Warm-blooded.

Extinct Referring to species with no living individuals.

Extirpate To eliminate a species from a particular region; local extinction.

Family (of lizards or crocodilians) A taxonomic group containing two or more closely related genera.

Fossorial Living underground.

Generalist An animal that does not specialize in any particular type of prey or is not restricted to a particular habitat.

Genus (*pl.* genera) A taxonomic grouping of one or more closely related species.

Herpetofauna The species of amphibians and reptiles that inhabit a given area.

Herpetologist A scientist who studies amphibians and reptiles.

Herpetology The study of amphibians and reptiles.

Hibernation A period of inactivity during cold periods; called "brumation" when referring to amphibians and reptiles.

Hybrid The offspring of mating between two different species.

Incubation period The time period between when eggs are laid and when they hatch.

Intergrade An intermediate form of a species resulting from mating and genetic mixing between individuals of two or more subspecies within a zone where their ranges overlap. Intergrade specimens may possess traits of all subspecies involved.

Nocturnal Active at night.

PARC Partners in Amphibian and Reptile Conservation; the largest partnership group dedicated to the conservation of all amphibians and reptiles and their habitats.

Parthenogenetic Capable of producing viable eggs without being fertilized by a male.

Pheromone A chemical that is released by an animal and used as a signal to other animals of the same species.

Radiotelemetry A method of tracking an animal's movements using a radiotransmitter attached to or implanted in an animal and a directional antenna and radio receiver.

Sandhills A habitat type in the southeastern Coastal Plain characterized by sandy soils, rolling topography, scrub oak, and longleaf or slash pine.

Scrub habitat Dry, sandy ridges of central and southern Florida with well-drained soils that are poor in nutrients. Vegetation in scrub habitat is characteristically an open or closed canopy of sand pines with an understory of dense clumps of shrubs, such as various scrub oaks, rosemary, and saw palmetto, separated by patches of barren sand.

Smooth scale A scale that is completely flat and has no ridge down the center.

Specialist An animal restricted to a particular diet or habitat.

Species Typically, an identifiable and distinct group of organisms composed of individuals capable of interbreeding and producing viable offspring under natural conditions.

Subspecies A taxonomic unit within a species, usually defined as morphologically distinct and occupying a geographic range that does not overlap with those of other such "races" of the species. Subspecies may interbreed naturally in areas of geographic contact. *See also* Intergrade.

Taxonomy The scientific field of classification and naming of organisms.

Warm-blooded A nontechnical term that refers to an animal that maintains its body temperature primarily with metabolic heat. *See also* Endotherm.

Further Reading

Ashton, R. E., Jr., and P. S. Ashton. 1985. *Handbook of Reptiles and Amphibians of Florida. Part II: Lizards, Turtles, and Crocodilians.* Miami, Fla.: Windward Publishing.

Bartlett, R. D., and P. P. Bartlett. 1999. *A Field Guide to Florida Reptiles and Amphibians.* Houston, Tex.: Gulf Publishing Company.

Behler, J. L., and F. W. King. 1979. *The Audubon Society Field Guide to North American Reptiles and Amphibians.* New York: Alfred A. Knopf.

Carmichael, P., and W. Williams. 2001. *Florida's Fabulous Reptiles and Amphibians.* Tampa, Fla.: World Publications.

Conant, R., and J. T. Collins. 1991. *A Field Guide to Reptiles and Amphibians of Eastern and Central North America.* 3rd ed. Boston: Houghton Mifflin.

Dundee, H. A., and D. A. Rossman. 1989. *The Amphibians and Reptiles of Louisiana.* Baton Rouge: Louisiana State University Press.

Gibbons, J. W., and R. D. Semlitsch. 1991. *Guide to the Reptiles and Amphibians of the Savannah River Site.* Athens: University of Georgia Press.

Jensen, J., C. Camp, J. W. Gibbons, and M. Elliott. 2008. *Reptiles and Amphibians of Georgia.* Athens: University of Georgia Press.

Lohoefener, R., and R. Altig. 1983. *Mississippi Herpetology.* Mississippi State University Research Center Bulletin 1:1–66.

Martof, B. S., W. M. Palmer, J. R. Bailey, J. R. Harrison III, and J. Dermid. 1980. *Amphibians and Reptiles of the Carolinas and Virginia.* Chapel Hill: University of North Carolina Press.

Meshaka, W. E., Jr., B. P. Butterfield, and J. B. Hauge. 2005. *Exotic Amphibians and Reptiles of Florida.* Melbourne, Fla.: Krieger Publishing.

Mitchell, J. C. 1994. *The Reptiles of Virginia.* Washington, D.C.: Smithsonian Institution Press.

Mount, R. H. 1975. *The Reptiles and Amphibians of Alabama.* Auburn, Ala.: Auburn University Agricultural Experiment Station.

Palmer, W. M., and A. L. Braswell. 1995. *Reptiles of North Carolina.* Chapel Hill: University of North Carolina Press.

Pianka, E., and L. Vitt. 2003. *Lizards: Windows to the Evolution of Diversity.* Berkeley: University of California Press.

Zim, H. S., and H. M. Smith. 2001. *Reptiles and Amphibians.* A Golden Guide. New York: St. Martin's Press.

Acknowledgments

We are grateful to the numerous staff and students at the Savannah River Ecology Laboratory for their encouragement and assistance in writing the book. In particular, we thank Margaret Wead for her help in preparing digital images of slides and assisting in a variety of other ways with the details of preparation. Teresa Carroll was of special importance to us in organization during the early stages. Among the students at SREL we thank Kimberly Andrews, Aaliyah Greene, Tom Luhring, Brian Todd, Tracey Tuberville, J. D. Willson, and Chris Winne, who along with Cris Hagen, David Scott, and Kurt Buhlmann were helpful in making suggestions and providing insights from their own experiences with lizards, alligators, and American crocodiles. Michael Dorcas was an excellent source of advice in preparation of the manuscript and selection of photographs.

We thank the following lizard and crocodilian specialists for their willingness to provide comments on individual species accounts, although we assume all responsibility for the final content: Ray Ashton, Mark Bailey, Dick Bartlett, Jeff Beane, Steve Bennett, Laura Brandt, Alvin Braswell, Carlos Camp, Todd Campbell, Bill Cooper, Brian Crother, Michael Dorcas, Kevin Enge, Greg Greer, Craig Guyer, Brian Halstead, Julian Harrison, John Jensen, Tom Jenssen, Bob Jones, Fred Kraus, Kenney Krysko, John MacGregor, Barry Mansell, Ken Marion, Frank Mazzotti, Earl McCoy, Bruce Means, Walter Meshaka, Joe Mitchell, Paul Moler, Tom Murphy, Nathan Parker, David Pike, Steve Price, Travis Robbins, Perran Ross, Floyd Scott, Stan Trauth, Laurie Vitt, J. D. Willson, Chris Winne, and Cameron Young.

We appreciate the excellent lizard and crocodilian photographs we received from colleagues willing to help the project. We are particularly grateful to several individuals for special assistance with photographs, questions about lizard or crocodilian biology, and preparation of maps. These include Berkeley Boone, Kevin Enge, Tony Gamble, Steve Godley, Greg Greer, Cris Hagen, Doug Hill, Mindy Hill, Kenney Krysko, Hayley McLeod, Paul Moler, and Scott Pfaff.

Credits

The authors would like to thank the following individuals and organizations for providing photographs:

Kimberly Andrews
Photograph on page 50 (middle right).

Craig Barrow III
Photograph on page 30 (bottom).

Richard Bartlett
Photographs on pages vii, 6 (left), 14 (top right), 15, 22 (top and bottom), 23, 25 (bottom), 30 (top), 36 (right), 72, 73, 85, 88, 90, 93, 95, 96 (upper middle, lower middle, and bottom), 100, 102, 105, 112, 113, 115, 121 (top), 128 (top), 130 (top), 132, 133 (top), 135 (top), 136 (top), 139 (top), 142, 146, 147, 148, 149, 151, 153, 154, 156, 157, 159 (top and bottom), 162 (top and bottom), 164 (top), 168 (bottom), 170, 172, 173 (bottom), 174 (top), 175, 176, 178, 179, 184, and 187 (bottom second).

Bettina Baumgartner / Fotolia.com
Photograph on page 64 (top).

Giff Beaton
Photograph on page 96 (top).

John Bell / istockphoto.com
Photograph on page 19 (top).

Anne Boston
Photograph on page 31 (left).

Kurt Buhlmann
Photograph on page 206 (bottom).

Todd Campbell
Photograph on page 186.

Tony Campbell / istockphoto.com
Photograph on page 44 (bottom).

CE / Fotolia.com
Photograph on page 66.

Dainis Derics / istockphoto.com
Photograph on page 215.

Andrew Durso
Photographs on pages 31 (right), 65 (left), and 68 (bottom).

Kevin Enge
Photographs on pages 14 (bottom left), 42, 125 (top and bottom), 126 (top and bottom), 127, 128 (bottom), 130 (bottom), 131, 133 (bottom), 134 (left and right), 135 (bottom), 136 (bottom), 137 (bottom), 139 (bottom), 140, 143 (top and bottom), 144, 150, 152 (top), 165 (top and bottom), 180, 181, 182, 199, and 206 (top).

Tony Gamble
Photograph on page 167.

Jennifer Gibbons
Photograph on page 219.

Mike Gibbons
Photographs on pages 191 (bottom) and 217.

Whit Gibbons
Photographs on pages 33 (bottom right), 47, 209, 210, and 212–213.

Jon Gorr / istockphoto.com
Photograph on page 40.

Matt Greene
Photographs on pages 2 (bottom), 16 (bottom), 18 (top and bottom), 24 (top), 33 (bottom left), 34, 35, 43, 51 (top), 52, 53 (middle left), 76, and 218.

Greg Greer
Photographs on pages 2 (top), 5, 9 (top), 13, 19 (bottom), 21, 28, 36 (left), 39 (right), 68 (top), 101 (bottom), 109, 117, 121 (bottom), 123 (top), 169, 177, 185, 187 (top first, second, third, and fourth and bottom first and third) and 196.

Maria Gritcai / istockphoto.com
Photograph on page v.

Kevin Gryczan / istockphoto.com
Photograph on page 54.

Mehmet Salih Guler / istockphoto.com
Photograph on pages 118–119.

Jackie C. Guzy / Biological Research Associates
Photograph on page 155.

Cris Hagen
Photographs on pages 11 (top), 41, 51 (bottom), 53 (bottom), 191 (top), 192, 193 (bottom), 194, 197, 205 (top), 207, and 211.

Bess Harris
Photographs on pages 6 (right), 69 (top), and 70 (left).

Aubrey M. Heupel
Photograph on page 24 (bottom).

Douglas K. Hill
Photograph on page 69 (bottom).

Eric Isselée / lifeonwhite.com / istockphoto.com
Photographs on pages i, ii–iii (bottom), 188–189, and 202–203.

Terri Jenkins
Photographs on page 46 (top, middle, and bottom).

Cathy Keifer / istockphoto.com
Photograph on page 12.

Kenney Krysko
Photographs on pages 32, 158, 161 (top), 166, 171 (top and bottom), and 174 (bottom).

Stacey Lance
Photographs on page 50 (top and bottom).

Megan Lorenz / istockphoto.com
Photograph on pages ii–iii (top).

Tom Luhring
Photographs on pages 1, 7, 10, 38 (top), 44 (top), 53 (top and middle right), 67 (left and right), and 195.

Larry Lynch / istockphoto.com
Photograph on page 55.

John MacGregor
Photographs on pages 20 (bottom) and 183.

Barry Mansell
Photographs on pages 26, 29 (top), 61, 63, 76 (bottom), 83, 92 (top and bottom), 98, 108, 110, 116, and 173 (top).

Tony Mills
Photograph on page 193 (top).

Justin Oguni
Photographs on pages 14 (top left) and 79.

Sean Poppy
Photograph on page 50 (middle left).

Leigh Prather / Dreamstime.com
Photographs on pages vi and 17.

Dave Rodriguez / istockphoto.com
Photograph on pages 38–39 (bottom).

Savannah River Ecology Laboratory
Photographs on pages 8, 45, 49 (top, lower left, and lower right), and 200.

Doug Schneider / istockphoto.com
Photograph on page 208.

David Scott
Photographs on pages 4 (bottom) and 198.

Michael Thompson / Dreamstime.com
Photograph on page 48.

Laurie Vitt
Photographs on pages 4 (top), 25 (top), 29 (bottom left and right), 33 (top), 70 (right), 107 (top and bottom), 114, 122, 123 (bottom), 124, 152 (bottom), and 163.

Kent Vliet
Photograph on page 205 (bottom).

Drazen Vukelic / Dreamstime.com
Photograph on page viii.

J. D. Willson
Photographs on pages 9 (bottom), 11 (bottom), 16 (top), 27, 37 (left and right), 58–59, 64 (bottom), 65 (right), 71 (left and right), 74, 75, 78, 80, 81, 82, 86, 89, 99 (top and bottom), 101 (top), 103, 104, 105 (top), 129, 137 (top), 138, 141 (top and bottom), 145 (top and bottom), 160, 161 (bottom), 164 (bottom), 168 (top), and 214.

Arkadiy Yarmolenko / istockphoto.com
Photograph on page 20 (top).

Index of Scientific Names

Boldface page numbers refer to species accounts. *Italicized* page numbers refer to illustrations.

Agama agama, **177**

Agama agama africana, 177

Alligator mississippiensis, **191–197**

Ameiva ameiva, **154**

Anguis fragilis, 80

Anolis carolinensis, **64–67**

Anolis carolinensis carolinensis, 65

Anolis carolinensis seminolus, 65

Anolis chlorocyanus, **125**

Anolis cristatellus, **126–127**

Anolis cristatellus cristatellus, 126

Anolis cybotes, **128–129**

Anolis cybotes cybotes, 128

Anolis distichus, **130–131**

Anolis distichus floridanus, 131

Anolis equestris, **132–133**

Anolis equestris equestris, 132

Anolis garmani, **134**

Anolis porcatus, **135–136**

Anolis sagrei, **137–139**

Anolis sagrei ordinatus, 137–138

Anolis sagrei sagrei, 137–138

Anolis sagrei stejnegeri, 138

Aspidoscelis motaguae, **153**

Aspidoscelis sexlineatus, **75–77**

Aspidoscelis sexlineatus sexlineatus, 75

Aspidoscelis sexlineatus stephensae, 75

Aspidoscelis sexlineatus viridus, 75

Basiliscus vittatus, **145–146**

Caiman crocodylus, **205–206**

Calotes mystaceus, **178**

Calotes veriscolor, **179**

Calotes versicolor versicolor, 179

Chamaeleo calyptratus, **181–182**

Chondrodactylus bibronii, **170**

Cnemidophorus lemniscatus, **151–152**

Cnemidophorus motaguae. See *Aspidoscelis motaguae*

Cnemidophorus sexlineatus. See *Aspidoscelis sexlineatus*

Cosymbotus platyurus. See *Hemidactylus platyurus*

Crocodylus acutus, **198–201**

Crotaphytus collaris, **121–122**

Ctenonotus cybotes. See *Anolis cybotes*

Ctenonotus distichus. See *Anolis distichus*

Ctenosaura pectinata, **142**

Ctenosaura similis, 142, **143–144**

Cyclura lewisi, 216

Draco volans, 38

Eublepharis macularius, 187

Eumeces. See individual species names in the genus Plestiodon

Gekko gecko, **157–158**

Gonatodes albogularis, **159**

Gonatodes albogularis fuscus, 159

Gonatodes albogularis notatus, 159

Heloderma suspectum, 117

Hemidactylus frenatus, **160–161**

Hemidactylus garnotii, **162–163**

Hemidactylus mabouia, **164–165**

Hemidactylus platyurus, **166–167**

Hemidactylus turcicus, **168–169**

Hemidactylus turcicus turcicus, 168

Iguana iguana, **140–141**

Leiocephalus carinatus, **147–148**

Leiocephalus carinatus armouri, 147

Leiocephalus carinatus coryi, 147

Leiocephalus carinatus virescens, 147–148

Leiocephalus schreibersi, **149–150**

Leiolepis belliana, **180**

Leiolepis belliana belliana, 180

Mabuya multifasciata, **156**

Mabuya perrotetii, 187

Neoseps reynoldsi. See *Plestiodon reynoldsi*

Norops garmani. See *Anolis garmani*

Norops sagrei. See *Anolis sagrei*

Ophisaurus attenuatus, **78–80**

Ophisaurus attenuatus attenuatus, 78, 79

Ophisaurus attenuatus longicaudus, 78, 79

Ophisaurus compressus, **81–82**

Ophisaurus mimicus, **83–85**

Ophisaurus ventralis, **86–88**

Pachydactylus bibronii. See *Chondrodactylus bibronii*

Phelsuma madagascariensis, **171–172**

Phelsuma madagascariensis madagascariensis, 171

Phrynosoma cornutum, **123–124**

Plestiodon anthracinus, **92–94**

Plestiodon anthracinus anthracinus, 92

Plestiodon anthracinus pluvialis, 92

Plestiodon egregius, **95–98**

Plestiodon egregius egregius, 95, 97

Plestiodon egregius insularis, 95, 97

Plestiodon egregius lividus, 95, 97

Plestiodon egregius onocrepis, 95, 97

Plestiodon egregius similis, 95, 97
Plestiodon fasciatus, **99–101**
Plestiodon inexpectatus,
    102–104
Plestiodon laticeps, **105–109**
Plestiodon reynoldsi, **110–112**
Plestiodon septentrionalis,
    113–114
*Plestiodon septentrionalis
    obtusirostris*, 113
Podarcis muralis, **183–184**
Pogona vitticeps, 187

Rhineura floridana, **61–63**

Sceloporus malachiticus, 187
Sceloporus undulatus, **68–71**
Sceloporus undulatus hyacinthinus,
    69
Sceloporus undulatus undulatus, 69
Sceloporus woodi, **72–74**
Scincella lateralis, **89–91**
Sphaerodactylus argus, **173**
Sphaerodactylus argus argus, 173
Sphaerodactylus elegans, **174–175**
Sphaerodactylus elegans elegans, 174

Sphaerodactylus notatus, **115–117**
Sphaerodactylus notatus notatus, 115

Tarentola annularis, **176**
Tarentola mauritanica, 187
Tiliqua scincoides, 187
Tupinambis merianae, **155**

Varanus exanthematicus, 187
Varanus niloticus, **185–186**
Varanus niloticus niloticus, 185
Varanus niloticus ornatus, 185

Index of Common Names

Boldface page numbers refer to species accounts. *Italicized* page numbers refer to illustrations.

African rainbow lizard, *2, 19,* **177**

agama, common. *See* African rainbow lizard

agama, red-headed. *See* African rainbow lizard

agama, red-headed rock. *See* African rainbow lizard

agama, tree. *See* Indochinese bloodsucker

alligator, 41–43; and defense mechanisms, 51; food and feeding of, 45, 47; habitat of, 52; and reproduction, 47–48; and research, 210–211

alligator, American, *1, 9,* 41–45, *52, 55,* **191–197**, *210, 217*; and conservation issues, 218, 219; and defense mechanisms, 51; food and feeding of, 45, *46,* 47; habitat of, *53*; identification of, *198, 206*; as pets, 213; and reproduction, 47–48, *49*; and research, 211, 212

alligator, Chinese, 41, *41*

alligator lizard, 16; northern, 29

ameiva, giant, *15,* **154**

anole, 12, 14–15, 25, 34, 38; as pet, 215; and research, 212

anole, Bahamian brown. *See* brown anole

anole, bark, **130–131**, 136; Florida, 131; green, 131

anole, blue-green. *See* anole, Hispaniolan green

anole, brown. *See* brown anole

anole, crested, **126–127**

anole, Cuban. *See* brown anole

anole, Cuban brown. *See* brown anole

anole, Cuban green, 66, 131, **135–136**

anole, Cuban knight. *See* knight anole

anole, green. *See* green anole

anole, Haitian green. *See* anole, Hispaniolan green

anole, Hispaniolan green, **125**

anole, Jamaican giant, *6,* **134**, 138

anole, knight. *See* knight anole

anole, large-headed, **128–129**

anole, Puerto Rican crested. *See* anole, crested

anole, Stejneger's. *See* brown anole

Argentine black-and-white tegu. *See* Argentine giant tegu

Argentine giant tegu, **155**

armadillo lizard, 150

bamboo chicken. *See* iguana, green

basilisk, 14

basilisk, brown, *5, 32,* **145–146**

basilisk, striped. *See* basilisk, brown

beaded lizard, 11, 23

bearded dragon, *187*

bloodsucker, Indochinese, **178**

bloodsucker, variable, **179**

blue-tongue skink, *187*

Brazilian "two-headed snake," 62

broad-headed skink, *2, 11, 18, 39,* 99, **105–109**, *209, 219*; food and feeding of, 27, 34; identification of, 36, 38, 39, 76, 101, 102

brown anole, *5, 6, 22, 26,* **137–139**; identification of, 65–66; predators of, 67, 147, 148

butterfly lizard, **180**

caiman, 41–43, *192, 198*; and defense mechanisms, 51; food and feeding of, 45, 47; habitat of, 52; as pets, 213; and reproduction, 47–48; and research, 210–211, 212

caiman, black, 206

caiman, brown. *See* caiman, spectacled

caiman, spectacled, *5, 42, 42,* **205–206**

caiman lizard, 23

California legless lizard, 16

chameleon, true, 12, 20

chameleon, Yemen. *See* veiled chameleon

chameleo verde. *See* anole, Cuban green

chipojo. *See* knight anole

chitchat. *See* gecko, common house

chuckwalla, 27

coal skink, *18,* **92–94**; identification of, 101, 102, 107, 113; northern, 92–93; southern, 92–93

collared lizard, *4, 5, 12, 14,* **121–122**

crocodile, 41–43; and defense mechanisms, 51; food and feeding of, 45, 47; habitat of, 52; and reproduction, 47–48; and research, 210–211, 212

crocodile, American, *4, 42, 42,* **198–201**; activity and locomotion of, 44; and conservation issues, 219; and defense mechanisms, 51; habitat of, 52; and reproduction, 48; and similarities to American alligator, 192

crocodile, dwarf, 41

crocodile, Nile, 195
crocodile, saltwater, 45, 47, 52, 195
Cuban green anole, 66, 131, **135–136**
curly-tailed lizard, Hispaniolan. *See* curly-tailed lizard, red-sided
curly-tailed lizard, northern, **147–148**
curly-tailed lizard, red-sided, **149–150**

emerald swift, *187*

fat-tailed gecko, *187*
fence lizard, eastern, 12, 14, 25, *30*, 34, 35, **68–71**; and Florida scrub lizard, **72**; identification of, 37, 38; northern, 69; southern, 69
field streak. *See* racerunner, six-lined
fire-sided skink, *187*
five-lined skink, *29*, 35; identification of, 37, 38, 39, 76, 94, 97, **102**
flying dragon, 38

gecko, 11, 18–19, 23, 28, 39, 212
gecko, African house. *See* wood slave
gecko, Ameriafrican house. *See* wood slave
gecko, ashy, 117, 165, **174–175**
gecko, Asian flat-tailed house, **166–167**
gecko, Bibron's (thick-toed), **170**

gecko, common house, **160–161**, 165
gecko, Cuban ashy. *See* gecko, ashy
gecko, day. *See* Madagascar day gecko
gecko, fat-tailed, *187*
gecko, flat-tail. *See* gecko, Asian flat-tailed house
gecko, Florida reef. *See* reef gecko

gecko, giant day. *See* Madagascar day gecko
gecko, Indo-Pacific. *See* Indo-Pacific gecko
gecko, leopard, *214*
gecko, MacCleay's ashy. *See* gecko, ashy
gecko, Madagascar day. *See* Madagascar day gecko
gecko, Mediterranean. *See* Mediterranean gecko
gecko, Moorish, *187*
gecko, ocellated, 117, **173**
gecko, reef. *See* reef gecko
gecko, ringed wall, 167, 176
gecko, Southeast Asian flying (parachute gecko), *32*
gecko, tropical. *See* wood slave
gecko, tropical house. *See* gecko, common house; wood slave
gecko, Turkish. *See* Mediterranean gecko
gecko, twin-spotted. *See* gecko, ringed wall
gecko, white-spotted. *See* gecko, ringed wall
gecko, yellow-headed, **159**
Gila monster, 11, 23, 117
glass lizard, 12, 16–17, 24, 31, 35; identification of, 36, 37, 38, 39
glass lizard, eastern, *16*, 35, 77, 83–84, **86–88**
glass lizard, European, 16
glass lizard, island, *23*, 79, **81–82**, 84, 86–87
glass lizard, mimic, 79, 81, **83–85**, 86–87
glass lizard, slender, *36*, **78–80**, 79, 87; eastern, 78; identification of, 81, 83–84, 86; western, 78, 79
goanna, 21
green anole, *7, 10, 14, 22, 24, 27, 31, 38, 54*, **64–67**; activity and locomotion of, 25; food and feeding of, 28, 139; habitat of, 35;

hibernation of, 34; identification of, 37, 38, 128, 135–136; northern, 65; predators of, 139, 148; and reproduction, 28, 30; southern, 65
green anole, Cuban, 66, 131, **135–136**

horned lizard, 14, 32
horned toad. *See* Texas horned lizard

iguana, 14–15, *215*
iguana, black. *See* spinytail iguana: black
iguana, Central American spinytail. *See* spinytail iguana: black
iguana, Grand Cayman blue, 216
iguana, green, *viii, 14, 32*, **140–141**
iguana, marine, 14, 15
iguana, Old World, 14
Indochinese bloodsucker, **178**
Indo-Pacific gecko, 19, 30, *30*, **162–163**, 165

Jesus Christ lizard. *See* basilisk, brown
jungle runner. *See* ameiva, giant

knight anole, *14*, 34, **132–133**
Komodo dragon, 11, *11*, 186

leopard lizard, 122
lion lizard. *See* curly-tailed lizard, northern
little brown skink, 35, **89–91**; identification of, 36, 39, 97, 110; predators of, *33*, 34, 71

mabuya, brown, **156**
Madagascar day gecko, **171–172**
Mediterranean gecko, 5, 18, 25, 165, **168–169**; identification of, 160, 162

mole skink, *29, 35, 89,* **95–98**, 110; blue-tailed, *95, 96, 97,* 216; Cedar Key, *95, 96, 97;* Florida Keys, *95, 97;* northern, *95, 96, 97;* peninsula, *95, 96, 97*

money monitor. *See* Nile monitor

monitor, *12, 21, 23*

mosasaurs, *3*

Nile monitor, *21,* **185–186**, *192*

Old World lizards, *12, 19*

Oriental garden lizard. *See* bloodsucker, variable

pigmy short-horned lizard, *29*

prairie skink, **113–114**; southern, 113

racerunner, *11, 31, 38*

racerunner, six-lined, *9, 15, 31, 34, 37,* **75–77**; activity and locomotion of, *24;* eastern, *75;* food and feeding of, *28;* identification of, *37;* and introduced species, *152, 153, 154;* prairie, *75;* predators of, *34*

racerunner, Texas yellow-headed, 75

rainbow lizard. *See* whiptail, rainbow

reef gecko, *11, 25, 36,* **115–117**; identification of, *37, 38, 173*

sand runner. *See* racerunner, six-lined

sand skink, *36, 89,* 216; Florida, *23*

sand streak. *See* racerunner, six-lined

Savannah monitor, *187*

scrub lizard, *38*

scrub lizard, Florida, *69,* **72–74**

six-lined racerunner. *See* racerunner, six-lined

skink, *12, 17–18;* activity and locomotion of, *25;* blue-tailed, *32;* and defense mechanisms, *31–32;* food and feeding of, *28;* identification of, *38;* predators of, *34, 156;* and reproduction, *30;* and research, *212*

skink, black. *See* coal skink

skink, blue-tongue, *187*

skink, broad-headed. *See* broad-headed skink

skink, brown. *See* mabuya, brown

skink, coal. *See* coal skink

skink, common five-lined, **99–101**, *106–107, 109*

skink, fire-sided, *187*

skink, five-lined. *See* five-lined skink

skink, Florida sand, *24, 97,* **110–112**

skink, golden. *See* mabuya, brown

skink, ground. *See* little brown skink

skink, little brown. *See* little brown skink

skink, many-lined glass. *See* mabuya, brown

skink, mole. *See* mole skink

skink, sand. *See* sand skink

skink, southern coal, *35*

skink, sun. *See* mabuya, brown

slow worm, *80*

snake lizard, *23*

southeastern five-lined skink, *99,* 101, **102–104**, *106*

spectacled caiman. *See* caiman, spectacled

spinytail iguana, *28, 32;* black, *32,* 142, **143–144**; Mexican, *32,* 142

tegu, *15;* Argentine giant, **155**

Texas horned lizard, *5, 12, 39,* **123–124**

tokay gecko, *19,* **157–158**

tomistoma, long-snouted (false gharial), *41*

veiled chameleon, *12, 20,* **181–182**

wall lizard, *20;* common, *20,* **183–184**

whiptail, *15, 15, 28, 30, 76, 77*

whiptail, Central American. *See* whiptail, giant

whiptail, giant, **153**

whiptail, rainbow, *28,* **151–152**

wood slave, **164–165**

wormlizard, *11, 13, 23;* Florida, *13, 24, 36,* **61–63**